NANOSTRUCTURES, NANOMATERIALS, AND NANOTECHNOLOGIES TO NANOINDUSTRY

NANOSTRUCTURES, NANOMATERIALS, AND NANOTECHNOLOGIES TO NANOINDUSTRY

Edited by
Vladimir I. Kodolov, DSc, Gennedy E. Zaikov, DSc, and A. K. Haghi, PhD

Apple Academic Press

TORONTO   NEW JERSEY

Apple Academic Press Inc. | Apple Academic Press Inc.
3333 Mistwell Crescent | 9 Spinnaker Way
Oakville, ON L6L 0A2 | Waretown, NJ 08758
Canada | USA

©2015 by Apple Academic Press, Inc.

First issued in paperback 2021

Exclusive worldwide distribution by CRC Press, a member of Taylor & Francis Group
No claim to original U.S. Government works

ISBN 13: 978-1-77463-313-7 (pbk)
ISBN 13: 978-1-926895-88-8 (hbk)

Library of Congress Control Number: 2014938963

Library and Archives Canada Cataloguing in Publication

Nanostructures, nanomaterials, and nanotechnologies to nanoindustry/edited by Vladimir I. Kodolov, DSc, Gennedy E. Zaikov, DSc, and A.K. Haghi, PhD.

(AAP research notes on nanoscience & nanotechnology book series)
Includes bibliographical references and index.
ISBN 978-1-926895-88-8 (bound)
1. Nanotechnology--Congresses. 2. Nanochemistry--Congresses. 3. Nanostructures--Congresses. 4. Nanostructured materials--Congresses. 5. Nanomedicine--Congresses. 6. Ultrastructure (Biology)--Congresses. I. Kodolov, Vladimir Ivanovich, author, editor II. Haghi, A. K., author, editor III. Zaikov, G. E. (Gennadiǐ Efremovich), 1935-, author, editor IV. International Conference "From Nanostructures, Nanomaterials and Nanotechnologies to Nanoindustry" (4th : 2013 : Izhevsk, Russia) V. Series: AAP research notes on nanoscience & nanotechnology book series

T174.7.N348 2014   620'.5   C2014-902833-4

Apple Academic Press also publishes its books in a variety of electronic formats. Some content that appears in print may not be available in electronic format. For information about Apple Academic Press products, visit our website at **www.appleacademicpress.com** and the CRC Press website at **www.crcpress.com**

ABOUT THE EDITORS

Vladimir I. Kodolov, DSc

Vladimir I. Kodolov, DSc, is Professor and Head of the Department of Chemistry and Chemical Technology at M. I. Kalashnikov Izhevsk State Technical University in Izhevsk, Russia, as well as Chief of Basic Research at the High Educational Center of Chemical Physics and Mesoscopy at the Udmurt Scientific Center, Ural Division at the Russian Academy of Sciences. He is also the Scientific Head of the Innovation Center at the Izhevsk Electromechanical Plant in Izhevsk, Russia.

Gennady E. Zaikov, DSc

Gennady E. Zaikov, DSc, is Head of the Polymer Division at the N. M. Emanuel Institute of Biochemical Physics, Russian Academy of Sciences, Moscow, Russia, and professor at Moscow State Academy of Fine Chemical Technology, Russia, as well as professor at Kazan National Research Technological University, Kazan, Russia. He is also a prolific author, researcher, and lecturer. He has received several awards for his work, including the Russian Federation Scholarship for Outstanding Scientists. He has been a member of many professional organizations and is on the editorial boards of many international science journals.

A. K. Haghi, PhD

A. K. Haghi, PhD, holds a BSc in urban and environmental engineering from University of North Carolina (USA); a MSc in mechanical engineering from North Carolina A&T State University (USA); a DEA in applied mechanics, acoustics and materials from Université de Technologie de Compiègne (France); and a PhD in engineering sciences from Université de Franche-Comté (France). He is the author and editor of 65 books as well as 1000 published papers in various journals and conference proceedings. Dr. Haghi has received several grants, consulted for a number of major corporations, and is a frequent speaker to national and international audiences. Since 1983, he served as a professor at several universities. He is currently Editor-in-Chief of the *International Journal of Chemoinformatics and Chemical Engineering* and *Polymers Research Journal* and on the editorial boards of many international journals. He is also a member of the Canadian Research and Development Center of Sciences and Cultures (CRDCSC), Montreal, Quebec, Canada.

ABOUT THE AAP RESEARCH NOTES ON NANOSCIENCE & NANOTECHNOLOGY BOOK SERIES

The AAP Research Notes on Nanoscience & Nanotechnology book series reports on research development in the field of nanoscience and nanotechnology for academic institutes and industrial sectors interested in advanced research.

Eli M. Pearce, PhD
Former President, American Chemical Society; Former Dean, Faculty of Science and Art, Brooklyn Polytechnic University, New York, USA

Mathew Sebastian, MD
Senior Consultant Surgeon, Elisabethinen Hospital, Klagenfurt, Austria; Austrian Association for Ayurveda

Charles Wilkie, PhD
Professor, Polymer and Organic Chemistry, Marquette University, Milwaukee, Wisconsin, USA

BOOKS IN THE AAP RESEARCH NOTES ON NANOSCIENCE & NANOTECHNOLOGY BOOK SERIES

Nanostructure, Nanosystems and Nanostructured Materials:
Theory, Production, and Development
Editors: P. M. Sivakumar, PhD, Vladimir I. Kodolov, DSc,
Gennady E. Zaikov, DSc,
A. K. Haghi, PhD

Nanostructures, Nanomaterials, and Nanotechnologies to Nanoindustry
Editors: Vladimir I. Kodolov, DSc, Gennady E. Zaikov, DSc, and
A. K. Haghi, PhD

Foundations of Nanotechnology:
Volume 1: Pore Size in Carbon-Based Nano-Adsorbents
A. K. Haghi, PhD, Sabu Thomas, PhD, and
Moein MehdiPour MirMahaleh

Foundations of Nanotechnology:
Volume 2: Nanoelements Formation and Interaction
Sabu Thomas, PhD, Saeedeh Rafiei, Shima Maghsoodlou, and
Arezo Afzali

CONTENTS

LIST OF CONTRIBUTORS

N. V. Andronova
Blokhin Russian Oncological Scientific Center, Russian Academy of Medical Science, Russia

R. Ansari
Department of Mechanical Engineering, University of Guilan, Rasht, Iran

L. A. Argatkina
Russia, Lead research institute of chemical technology, Kashirsky high road, 115409, Moscow

A. Yu. Bonda
Basic Research-High Educational Center of Chemical Physics and Mesoscopy, Udmurt Scientific Center, Russian Academy of Sciences; Izhevsk, Udmurt Republic, Russia; E-mail: kodol@istu.udm.ru

N. V. Bondareva
Izhevsk State Technical University

V. E. Bozhevolnov
Faculty of Chemistry of Lomonosov Moscow State UniversityMSU, Faculty of Chemistry, Russia, 119991, Moscow 1, GSP-1, 1-3 Leninskiye Gory, E-mail: nic@radio.chem msu.ru

E. A. Budovskikh
Siberian State Industrial University, Russian Academy of Sciences, 132 Kirov St., Izhevsk, 426000

S. G. Bystrov
Physical-Technical Institute, Ural Division, Russian Academy of Sciences, ul. Kirova 132, Izhevsk, 426001, Russia. Tel.: (3412) 72-87-79; E-mail: bystrov.sg@mail.ru

F. F. Chausov
Udmurt State University, 1 Universitetskaya St., Izhevsk, 426034

N. V. Dezhkunov
Belarusian State University of Informatics and Radioelectronics

K. S. Dibirova
Dagestan State Pedagogical University, Makhachkala 367003, Yaragskii st., 57, Russian Federation

I. V. Dolbin
Kabardino-Balkarian State University, KBR, Nal'chik 360004, Chernyshevskii st., 173, Russian Federation

E. L. Dzidziguri
Russia, National research technological university «MISIS», 119049, Moscow, Leninsky prospect, 4. E-mail: dmitrykrylov@rambler.ru

M. S. Emelyanova
Izhevsk State Technical University

A. Y. Fedotov
Institute of Mechanics, Ural Branch of the Russian Academy of Sciences Department "Mechanics of Nanostructures," Izhevsk State Technical University, Department "Nanotechnology and Microsystems," Izhevsk, Russia T. Baramsinoy 34, Izhevsk, Russia E-mail: postmaster@ntm.udm.ru

A. V. Gopin
Faculty of Chemistry of Lomonosov Moscow State University MSU, Faculty of Chemistry, Russia, 119991, Moscow 1, GSP-1, 1-3 Leninskiye Gory, E-mail: nic@radio.chem.msu.ru

V. Ye. Gromov
Siberian State Industrial University, Russian Academy of Sciences, 132 Kirov St., Izhevsk, 426000

B. Hadavi Moghadam
Department of Textile Engineering, University of Guilan, Rasht, Iran

A. K. Haghi
Department of Textile Engineering, University of Guilan, Rasht, Iran, E-mail: Haghi@Guilan.ac.ir

M. Hasanzadeh
Department of Textile Engineering, University of Guilan, Rasht, Iran

Y. S. Ilina
Bauman Moscow State Technical University

Yu. F. Ivanov
Institute of high-current electronics SB RAS

N. V. Khokhriakov
Izhevsk State Agricultural Academy, Basic Research and Educational Center of Chemical Physics and Mesoscopy, Udmurt Scientific Center, Ural Division, Russian Academy of Science, Russia, Izhevsk, 426000, E-mail: korablev@udm.net.

V. I. Kodolov
Basic Research – High Educational Centre of Chemical Physics and Mesoscopy, Udmurt Scientific Center, Ural Division, Russian Academy of Sciences, Izhevsk, Russia, M.T. Kalashnikov Izhevsk State Technical University, Izhevsk, 426067, Russia

G. A. Korablev
Izhevsk State Agricultural Academy, Basic Research and Educational Center of Chemical Physics and Mesoscopy, Udmurt Scientific Center, Ural Division, Russian Academy of Science, Russia, Izhevsk, 426000, E-mail: korablev@udm.net.

R. G. Korablev
Izhevsk State Agricultural Academy, Basic Research and Educational Center of Chemical Physics and Mesoscopy, Udmurt Scientific Center, Ural Division, Russian Academy of Science, Russia, Izhevsk, 426000

A. Korablev
Izhevsk State Agricultural Academy, Basic Research and Educational Center of Chemical Physics and Mesoscopy, Udmurt Scientific Center, Ural Division, Russian Academy of Science, Russia, Izhevsk, 426000

O. A. Kovyazina
Basic Research-High Educational Center of Chemical Physics and Mesoscopy, Udmurt Scientific Center, Russian Academy of Sciences; Izhevsk, Udmurt Republic, Russia; E-mail: kodol@istu.udm.ru

G. V. Kozlov
N. M. Emanuel Institute of Biochemical Physics of Russian Academy of Sciences, Moscow 119334, Kosygin st., 4, Russian Federation

D. A. Krylov
Bauman Moscow State Technical University

T. V. Kuznetsova
State National Research Politechnical University of Perm, Perm, Russia

M. I. Lebedeva
Russia, National research technological university "MISIS", Leninsky prospect, 4, 119049, Moscow

G. V. Lomaev
Izhevsk State Technical University

G. M. Magomedov
Dagestan State Pedagogical University, Makhachkala 367003, Yaragskii st., 57, Russian Federation

V. Mottaghitalab
Department of Textile Engineering, University of Guilan, Rasht, Iran

A. V. Murin
Physicotechnical Institute, Ural Division of the Russian Academy of Sciences, 132 Kirov St., Izhevsk, 42600, aleksey.v.murin@gmail.com

E. A. Naimushina
Udmurt State University, 1 Universitetskaya St., Izhevsk, 426034

A. L. Nikolaev
Faculty of Chemistry of Lomonosov Moscow State University MSU, Faculty of Chemistry, Russia, 119991, Moscow 1, GSP-1, 1-3 Leninskiye Gory, E-mail: nic@radio.chem.msu.ru

A. K. Osipov
Izhevsk State Agricultural Academy

N. G. Petrova
Ministry of Informatization and Communication of the Udmurt Republic

Ya. Polyotov
Scientific Education Centre of Chemical Physics and Mesoscopy, the Ural Branch of the Russian Academy of Sciences, T. Baramzina St., 34, Izhevsk, 426067.

S. V. Raykov
Siberian State Industrial University, Russian Academy of Sciences, 132 Kirov St., Izhevsk, 426000
D. A. Romanov
Siberian State Industrial University, Russian Academy of Sciences, 132 Kirov St., Izhevsk, 426000

A. V. Severyukhin
Institute of Mechanics, Ural Branch of the Russian Academy of Sciences Department "Mechanics of Nanostructures." Izhevsk State Technical University, Department "Nanotechnology and Microsystems," Izhevsk, Russia T. Baramsinoy 34, Izhevsk, Russia E-mail: postmaster@ntm.udm.ru

O. Yu. Severyukhina
Institute of Mechanics, Ural Branch of the Russian Academy of Sciences, Department "Mechanics of Nanostructures." Izhevsk State Technical University, Department "Nanotechnology and Microsystems," Izhevsk, Russia T. Baramsinoy 34, Izhevsk, Russia E-mail: postmaster@ntm.udm.ru

I. N. Shabanova
Physicotechnical Institute of the Ural Branch of the Russian Academy of Sciences, 132 Kirov St. , Izhevsk 426000, xps@fti.udm.ru, FGBOU GOU VPO "Udmurt State University," 1 Universitetskaya St., Izhevsk

I. N. Shabanova
Physicotechnical Institute of the Ural Department of the Russian Academy of Sciences, 132 Kirov St. , Izhevsk 426000, xps@fti.udm.ru, FGBOU GOU VPO "Udmurt State University," 1 Universitetskaya St., Izhevsk

O. I. Shavrin
Kalashnikov Izhevsk State Technical University

A. A. Shushkov
Institute of Mechanics, Ural Branch of the Russian Academy of Sciences Department "Mechanics of Nanostructures" Izhevsk State Technical University, Department "Nanotechnology and Microsystems," Izhevsk, Russia T. Baramsinoy 34, Izhevsk, Russia E-mail: postmaster@ntm.udm.ru

N. I. Sidnyaev
Bauman Moscow State Technical University

N. V. Somov
State University of Nizhny Novgorod, 23 prospect Gagarina, Nizhny Novgorod, 603950

G. Song
Advanced Materials Institute of Graduate School at Shenzhen, Tsinghua University, Tsinghua Campus, The University Town Shenzhen

G. Tang
Advanced Materials Institute of Graduate School at Shenzhen, Tsinghua University, Tsinghua Campus, The University Town Shenzhen

N. S. Terebova
Physicotechnical Institute of the Ural Branch of the Russian Academy of Sciences, 132 Kirov St. , Izhevsk 426000, xps@fti.udm.ru, FGBOU GOU VPO "Udmurt State University," 1 Universitetskaya St., Izhevsk

E. M. Treschalina
Blokhin Russian Oncological Scientific Center, Russian Academy of Medical Science

V. V. Trineeva
Basic Research – High Educational Centre of Chemical Physics and Mesoscopy, Udmurt Scientific Center, Ural Division, Russian Academy of Sciences, Izhevsk, Russia, Institute of Mechanics, Russian Academy of Sciences, Izhevsk, Russia

V. N. Trofimov
State National Research Politechnical University of Perm, Perm, Russia

A. V. Vakhrushev
Institute of Mechanics, Ural Branch of the Russian Academy of Sciences,Department "Mechanics of Nanostructures," Izhevsk State Technical University, Department "Nanotechnology and Microsystems," Izhevsk, Russia T. Baramsinoy 34, Izhevsk, Russia E-mail: postmaster@ntm.udm.ru

Yu. M. Vasil'chenko
Basic Research – High Educational Center of Chemical Physics and Mesoscopy, Udmurt Scientific Center, Ural Division, Russian Academy of Sciences, Russia, M.T. Kalashnikov Izhevsk State Technical University, Russia

Yu. G. Vasiliev
Izhevsk State Agricultural Academy, Basic Research and Educational Center of Chemical Physics and Mesoscopy, Udmurt Scientific Center, Ural Division, Russian Academy of Science, Russia, Izhevsk, 426000, E-mail: korablev@udm.net.

G. E. Zaikov
N. M. Emanuel Institute of Biochemical Physics, RAS, Russia, Moscow, 119991, 4 Kosygin St., E-mail: chembio@chph.ras.ru

LIST OF ABBREVIATIONS

AFM	Atomic Force Microscopy
CFM	Chemical Force Microscopy
CNT	Carbon Nanotube
CSR	Coherent-Scattering Regions
DFT	Density Functional Theory
EAM	Embedded-Atom Method
EBP	Electron-Beam Processing
EEA	ElectroExplosive Alloying
EM	Electron Microscopy
EMI	Electromagnetic Interface
ERM	Extended Rule of Mixtures
eV	Electron-Volts
FDS	Fine-Dispersed Suspension
FE	Finite Element
FSM	Fourier Spectrophotometer
HDPE	High Density Polyethylene
HTMT	High-Temperature Thermomechanical Treatment
LDA	Local Density Approximation
LFE	Linear Dependencies of Free Energies
LnOF	Leads to Oxyfluoride Phase
MC	Monte Carlo
MD	Molecular Dynamics
Me	Metal
MF	Magnetic Fields
MP	Molding Apparate
NC	Nanocomposites
PA	Polyamide
PCM	Polymeric Composite Materials
PEPA	Polyethylene Polyamine
PMMA	Polymethylmethacrylate

PP	Polypropylene
PS	Polystyrene
PVA	Polyvinyl Alcohol
RDF	Radial Distribution Function
REE	Rare Earth Elements
RR	Ray-Ran
RVE	Representative Volume Element
SCRs	Selective Chemical Reactions
SEP	Spatial-Energy Parameter
SPSs	Solid-Phase Sonosensitizers
SSTMT	Small-Strain Thermomechanical Treatment
SWNTs	Single-Walled Nanotubes
TBMD	Tight Bonding Molecular Dynamics
TFAA	Trifluoroacetic Anhydride
TS	Test Samples
XPS	X-ray Photoelectronic Spectroscop
ZnPc	Zinc Octacarboxyphthalocyanine

LIST OF SYMBOLS

E_∞	applied external electric field
τ_{ZZ}	axial viscous normal stress
λ	conduction coefficient
Q	constant volume flow rate
$\bar{\varepsilon}$	dielectric constants of the jet
ρ	fluid density
σ_x	longitudinal pressure
m_i	mass of the i-th atom
E	modulus of elasticity
t_i^e	normal tractions
K	number of atoms
v	poisson's ratio
u	radial displacements
r_0	radius of the sphere
σ_s	radius of the spinneret
σ_s	resistance to deformation
u^*	temperature of the phase transition
c	volume fraction of the nanoparticles
a	nanoreactor activity
a_2	integers
b	index connected with the process rate
c1s	spectrum of carbon copper
d	dimension of Euclidean space
d_f	dimension of polymer structure
d_m	diameter of gas-penetrant
d_p	nanoparticle diameter

E	axial component of the electric field
e	elasticity modulus
e_b	binding energy
e_m	elasticity moduli
e^t	elasticity modulus
E_t	normal and tangential components
F	area of section of a profile
f	faraday number
F_1, F_2	total area of section of sectors
f_n	friction coefficient
g	acceleration due to gravity
g_{am}	shear modulus of amorphous phase
h	film thickness
I	constant total current in the jet
k	Boltzmann's constant
K	electrical conductivity of the liquid jet
k	process balance constant
k_{nc}	constant of coagulation rate of nanocomposite
k_{nt}	constant of coagulation rate of carbon nanotube
m	nanostructure mass
m_e	molecular weight of polymer chain
n	bond average repetition factor
n	number reflecting the charge of chemical particles
n_a	Avogadro number
n_p	number of moles of the product
n_r	number of moles of reagents
p	pressure
p_m	matrix polymer
p_n	permeability to gas coefficient
Q	heat of the phase transition
r	radius of rotating bodies

R'	slope of the jet surface
s	surface of nanoreactor walls
sp	satellite
t	process temperature
t	time
u	for temperature
V	control volume
v	nanoreactor volume
w	share of nanoproduct
w_n	mass contents
z	number of electrons
α	angle between the vector direction of the magnetic moment
$\alpha_{am,\,n}$	relative fractions of amorphous phase
γ	surface tension
δf	isothermal potential
$\delta\varphi$	difference of potentials at the boundary
$\varepsilon^0_s d$	multiplication of surface layer energy
ε^0_v	energy of nanoreactor volume unit
ε_s	surface energy
ε_v	nanoreactor volume energy
η	fluid viscosity
Θ	angle of diffraction
ϑ_{max}	heat release maximum rate
μ_0	magnetic constant
ν	oscillation frequency
ν	Poisson's ratio
$\nu_{кc}$	number of chemical bonds changing during the process
$\nu_{нc}$	number of initial state of chemical bonds
π	electrons
ρ_n	nanofiller density
ρ_p	polymer density
σ	surface charge density

τ	duration
υ	crystallinity degree
υ_κ	velocity of nanostructure oscillations
φ	suspension twist angle
K	hybridization coefficient
M	magnetic moment
H	magnetic field intensity
P_S	structural parameter

PREFACE

This volume accumulates the most important information about new trends in the nanochemistry and nanotechnology, and also in the nanobiology and nanomedicine field. The book contains the papers prepared based on the lectures presented at the Fourth International Conference "From Nanostructures, Nanomaterials and Nanotechnologies to Nanoindustry," which took place in Izhevsk, Russia, on April 3–5, 2013. The conference scientific program covers the production of nanostructures and nanosystems, nanomaterial science, investigation of nanostructures and nanostructured materials, development of prognostic apparatus for synthesis and investigation of nanoproducts, as well as the application of nanoproducts and nanotechnologies in different fields.

The editors selected papers, including reviewed articles, in the different fields of nanochemistry, nanobiology, nanomedicine, nanotechnology, computer modeling of nanosystems and their obtaining processes.

The book comprises information on the following trends:

- Nanostructures and nanosystems—theory and synthesis.
- The processes modeling and computer simulation.
- Surface investigations.
- Investigation methods development for nanosystems and nanostructured materials.
- Nanostructured materials properties and applications.
- Nanostructures and nanosystems in nanobiology and nanomedicine.

The book provides discussion on the following topics:

1. Perspectives of nanochemistry development for metal/carbon nanocomposites synthesis and for the material self-organization.
2. The metal/carbon nanocomposites synthesis in nanoreactors of polymeric matrixes.
3. Production and application of metal/carbon nanocomposites.

4. Calculation of the elastic parameters of composite materials based on nanoparticles using multilevel models.
5. Simulation of formation of superficial mono hetero structures.
6. Computer simulation of Ni, Cu and Au cluster structures in super cooled liquid state.
7. The calculating for metal/carbon nanocomposites synthesis in polymeric matrixes with the application of Avrami equations.
8. On diversified demonstration of entropy.
9. Research of the local physical and chemical structure of super thin layers of polymeric composite materials.
10. Structure and surface layer properties of medium carbon steel after electro explosive copper plating and subsequent electron-beam processing.
11. The structural analysis of nanocomposites polymer/organoclay flame-resistance.
12. The metal/carbon nanocomposites influence mechanism on media and on compositions.
13. X-ray study of the influence of the amount and activity of carbon metal containing nanostructures on the polymer modification.
14. Semi-crystalline polymers as natural hybrid nanocomposites: reinforcement degree.
15. Research of structure and polishing properties of nanopowders based on cerium dioxide.
16. Temperature field control and forecast in nanocompositional materials.
17. Influence of nanotechnology on coiled springs operational characteristics.
18. Drawing of continuous profiles of not round cross section for production of composite low-temperature superconductors.
19. Synthesis, structure and protective properties of tetra sodium nitrile-tris-methylene phosphonate zincate tridecahydrate $Na_4[N(CH_2PO_3)_3Zn]\cdot13H_2O$.
20. Influence of geomagnetic field variation of Fe concentration in the feed on feed mineralization in the Ontogenesis of *Apis mellifera* (L).
21. Transformation of high-energy bonds in ATP.

22. Ultrasonic nanomedicine in the aspect of cancer therapy.

23. A detailed review on production of electrospun CNT-polymer composite nanofibers.

This book is unique and important because the new trends in nano-chemistry, nanobiology, nanotechnology, nanomedicine as well as new objects of nanostructures, nanosystems and also nanostructured materials are discussed. The purchasers (researchers, professors, post gradients, students and other readers) will find many interesting information and receive new knowledge in the trend of nanoscience and nanotechnology.

The book will be useful for researchers, professors/instructors (for teaching specific courses), students and post-graduates and also for personal requalification, university/college libraries, and bookstores.

The editors and contributors will be happy to receive some comments from readers.

— **Vladimir I. Kodolov, DSc, Gennedy E. Zaikov, DSc,**
and A. K. Haghi, PhD

ABOUT THE CONFERENCE

From the 3rd to 5th of April 2013, the Kalashnikov Izhevsk State Technical University hosted the IV International Conference "From Nanostructures, Nanomaterials and Nanotechnologies to Nanoindustry."

The conference program was comprised of the presentations from Russian and other international scientists from the largest research centers. The scientific program covered obtaining and manufacturing of nanostructures, nanomaterial science, investigation of nanostructures and nanomaterials, development of prognostication apparatus when obtaining and investigating nanoproducts, as well as the application of nanoproducts and nanotechnologies in different areas.

The aim of the conference was discussion of the latest achievements in nanomaterial science and nanotechnology and the challenges in nanoindustry development.

Apart from plenary and oral presentations, there were computer poster presentations as well as a round-table discussion on the results of mastering nanotechnologies, semiindustrial and industrial production of nanocomposites and nanomaterials, and practical introduction of nanomaterials and nanotechnologies into different areas, including medicine and agriculture. The round-table discussion combined the reports of representatives from industrial enterprises and research institutions. The exhibition of modern equipment for the investigation of nanostructures, nanosystems and nanomaterials, as well as nanotechnological units and machinery of well-known instrument-makers, were also scheduled.

IV International Conference "From Nanostructures, Nanomaterials and Nanotechnologies to Nanoindustry" opened at 10:00 on April 03, 2013 at Student Palace "Integral." After the official welcoming addressed, the leading scientists in nanotechnologies made the presentations.

PART I

NANOSTRUCTURES AND NANOSYSTEMS THEORY AND SYNTHESIS

CHAPTER 1

PERSPECTIVES OF NANOCHEMISTRY DEVELOPMENT FOR METAL/CARBON NANOCOMPOSITES SYNTHESIS AND FOR THE MATERIALS SELF ORGANIZATION

V. I. KODOLOV and V. V. TRINEEVA

CONTENTS

ABSTRACT

This chapter is dedicated to the development of Nanochemistry methods for the Metal/Carbon Nanocomposites Synthesis as well as for the Materials modification by these Nanocomposites. The perspectives of the scientific trend introducing with the organization of modern Nanocomposites production in Nanoindustry are discussed. Nanochemistry methods for the creation of Metal/Carbon Nanocomposites in nanoreactors of polymeric matrixes are considered. The principal characteristics of nanocomposites obtained are given. With the help of IR and X-ray photoelectron spectroscopies it is found that the media "respond" to the introduction of super small quantities of nanostructures. The results of the modification of inorganic and organic materials with super small quantities of fine dispersed suspensions of Metal/Carbon Nanocomposites are presented.

1.1   INTRODUCTION

At last time the option exists that the basis of nanotechnology is self-organization of systems [1].

System self-organization refers to synergetics [2]. Quite often, especially recently, the papers are published, for example, by Malinetsky [3], in which it is considered that nanotechnology is based on self-organization of metastable systems. As assumed [4], self-organization can proceed by dissipative (synergetic) and continual (conservative) mechanisms. At the same time, the system can be arranged due to the formation of new stable ("strengthening") phases or due to the growth provision of the existing basic phase. This phenomenon underlies the arising nanochemistry.

Assuming that nanoparticle oscillation energies correlate with their dimensions and comparing this energy with the corresponding region of electromagnetic waves, we can assert that energy action of nanostructures is within the energy region of chemical reactions. Therefore one of the possible definitions of nanochemistry can be following.

Nanochemistry is a Science Investigating Nanostructures and Nanosystems in Metastable ("Transition") States and Processes Flowing with them in Near-"Transition" State or in "Transition" State with Low Activation Energies.

To carry out the processes based on the notions of nanochemistry, the directed energy action on the system is required, with the help of chemical particle field as well, for the transition from the prepared near-"transition" state into the process product state (in our case – into nanostructures or nanocomposites). The perspective area of nanochemistry is the chemistry in nanoreactors. Nanoreactors can be compared with specific nanostructures representing limited space regions in which chemical particles orientate creating "transition state" prior to the formation of the desired nanoproduct. Nanorectors have a definite activity, which predetermines the creation of the corresponding product. When nanosized particles are formed in nanoreactors, their shape and dimensions can be the reflection of shape and dimensions of the nanoreactor [5].

1.2 THEORETICAL PREMISES FOR OBTAINING AND APPLICATION OF THE NANOCOMPOSITES

Previously, we [5] proposed the parameter called the nanosized interval (B), in which the nanostructures demonstrate their activity. Depending on the structure and composition of nanoreactor internal walls, distance between them, shape and size of nanoreactor, the nanostructures differing in activity are formed. The correlation between surface energy, taking into account the thickness of surface layer, and volume energy was proposed as a measure of the activity of nanostructures, nanoreactors and nanosystems [1].

In this case, we obtain the absolute dimensionless characteristic (a) of nanostructure or nanoreactor activity.

$$a = \varepsilon_s^\circ \, d/\varepsilon_V^\circ \cdot N/r(h) = \varepsilon_s^\circ \, d/\varepsilon_V^\circ \cdot 1/B, \tag{1}$$

where B equals r(h)/N, R – radius of rotating bodies, including the hollow ones, h – film thickness, n – number changing depending on the nanostructure shape. Parameter d characterizes the nanostructure surface layer thickness, and corresponding energies of surface unit and volume unit are defined by the nanostructure composition.

The proposed scheme of obtaining carbon/metal-containing nano-structures in nanoreactors of polymeric matrixes includes the selection of polymeric matrixes containing functional groups. 3d metals (iron, cobalt, nickel, copper) are selected as the elements coordinating functional groups. The elements indicated easily coordinate with functional groups containing oxygen, nitrogen, halogens. Depending on metal coordinating ability and conditions for nanostructure obtaining (in liquid or solid medium with minimal content of liquid) we obtain "embryos" of future nano-structures of different shapes, dimensions and composition. It is advisable to model coordination processes and further redox processes with the help of quantum chemistry apparatus, following step-by-step consideration in accordance with the planned scheme.

At the same time, the metal orientation proceeds in interface regions and nanopores of polymeric phase which conditions further direction of the process to the formation of metal/carbon nanocomposite. In other words, the birth and growth of nanosize structures occur during the process in the same way as known from the macromolecule physics [6], in which Avrami equations are successfully used. The application of Avrami equations to the processes of nanostructure formation was previously discussed in the papers dedicated to the formation of ordered shapes of macromolecules [6], formation of carbon nanostructures by electric arc method [7], obtaining of fiber materials [8].

As follows from Avrami equation,

$$1-\upsilon = \exp[-k\tau^n], \tag{2}$$

where υ – crystallinity degree, τ – duration, k – value corresponding to specific process rate, n – number of degrees of freedom changing from 1 to 6, the factor under the exponential is connected with the process rate with the duration (time) of the process. Under the conditions of the isothermal growth of the ordered system "embryo," it can be accepted that the nanoreactor activity will be proportional to the process rate in relation to the flowing process. Then the share of the product being formed (W) in nanoreactor will be expressed by the following equation:

$$W = 1 - \exp(-a\tau^n) = 1 - \exp[-(\varepsilon_s/\varepsilon_v)\tau^n] = 1 - \exp\{-[(\varepsilon^\circ_s d/\varepsilon^\circ_v)S/V]\,\tau^n\}, \tag{3}$$

where a – nanoreactor activity, ε_s – surface energy reflecting the energy of interaction of reagents with nanoreactor walls, ε_v – nanoreactor volume energy, $\varepsilon^o_s d$ – multiplication of surface layer energy by its thickness, ε^o_v – energy of nanoreactor volume unit, S – surface of nanoreactor walls, V – nanoreactor volume.

When the metal ion moves inside the nanoreactor with redox interaction of ion (mol) with nanoreactor walls, the balance setting in the pair "metal containing – polymeric phase" can apparently be described with the following equation.

$$zF\Delta\varphi = RT\ln K = RT\ln(N_p/N_r) = RT\ln(1 - W), \qquad (4)$$

where z – number of electrons participating in the process; $\Delta\varphi$ – difference of potentials at the boundary "nanoreactor wall – reactive mixture"; F – Faraday number; R – universal gas constant; T – process temperature; k – process balance constant; N_p – number of moles of the product produced in nanoreactor; N_r – number of moles of reagents or atoms (ions) participating in the process which filled the nanoreactor; W – share of nanoproduct obtained in nanoreactor.

In turn, the share of the transformed components participating in phase interaction can be expressed with the equation, which can be considered as a modified Avrami equation.

$$W = 1 - \exp[-\tau^n \exp(zF\Delta\varphi/RT)], \qquad (5)$$

where τ – duration of the process in nanoreactor; n – number of degrees of freedom changing from 1 to 6.

During the redox process connected with the coordination process, the character of chemical bonds changes. Therefore correlations of wave numbers of the changing chemical bonds can be applied as the characteristic of the nanostructure formation process in nanoreactor.

$$W = 1 - \exp[-\tau^n(v_{HC}/v_{KC})], \qquad (6)$$

where v_{HC} corresponds to wave numbers of initial state of chemical bonds, and v_{KC} – wave numbers of chemical bonds changing during the process.

Modified Avraami equations were tested to prognosticate the duration of the processes of obtaining metal/carbon nanofilms in the system "Cu – PVA" at 200°C [10]. The calculated time (2.5 h) correspond to the experimental duration of obtaining carbon nanofilms on copper clusters.

The nanostructures formed in nanoreactors of polymeric matrixes can be presented as oscillators with rather high oscillation frequency. It should be pointed out that according to references [1] for nanostructures (fullerenes and nanotubes) the absorption in the range of wave numbers 1300–1450 cm^{-1} is indicative. These values of wave numbers correspond to the frequencies in the range 3.9–4.35×10^{13} Hz, that is, in the range of ultrasound frequencies.

If the medium into which the nanostructure is placed blocks its translational or rotational motion giving the possibility only for the oscillatory motion, the nanostructure surface energy can be identified with the oscillatory energy.

$$\varepsilon_S \approx \varepsilon_{\kappa} = m \upsilon_{\kappa}^2/2, \qquad (7)$$

where m – nanostructure mass, a υ_{κ} – velocity of nanostructure oscillations. Knowing the nanostructure mass, its specific surface and having identified the surface energy, it is easy to find the velocity of nanostructure oscillations.

$$\upsilon_{\kappa} = \sqrt{2\varepsilon_{\kappa}/m} \qquad (8)$$

If only the nanostructure oscillations are preserved, it can be logically assumed that the amplitude of nanostructure oscillations should not exceed its linear nanosize, that is, $\lambda < r$. Then the frequency of nanostructure oscillations can be found as follows:

$$v_{\kappa} = \upsilon_{\kappa}/\lambda \qquad (9)$$

Therefore, the wave number can be calculated and compared with the experimental results obtained from IR spectra.

The influence of nanostructures on the media and compositions was discussed based on quantum-chemical modeling [9]. After comparing the energies of interaction of fullerene derivatives with water clusters, it was found that the increase in the interactions in water medium under the nanostructure influence is achieved only with the participation of hydroxyfullerene in the interaction. The energy changes reflect the oscillatory process with periodic boosts and attenuations of interactions. The modeling results can identify that the transfer of nanostructure influence onto the molecules in water medium is possible with the proximity or concordance of oscillations of chemical bonds in nanostructure and medium. The process of nanostructure influence onto media has an oscillatory character and is connected with a definite orientation of particles in the medium in the same way as reagents orientate in nanoreactors of polymeric matrixes. To describe this process, it is advisable to introduce such critical parameters as critical content of nanoparticles, critical time and critical temperature [10]. The growth of the number of nanoparticles (n) usually leads to the increase in the number of interaction (N). Also such situation is possible when with the increase of n critical value, n value gets much greater than the number of active nanoparticles. If the temperature exceeds the critical value, this results in the distortion of self-organization processes in the composition being modified and decrease in nanostructure influence onto media.

1.3 MATERIALS AND METHODS

1.3.1 MATERIALS AND METHODS OF METAL/CARBON NANOCOMPOSITES SYNTHESIS

Based on theoretical notions the synthesis scheme of metal/carbon nanocomposites comprises two stages. At the first stage, nanoreactors are prepared in selected polymeric matrixes and filled with metal containing phase. As polymeric matrixes it is proposed to use polyvinyl alcohol, polyvinyl acetate and polyvinyl chloride differing by crystallinity degree, correlation of functional groups, swelling degree and dimensions of interlayer spaces. Metal containing phase represents chlorides or oxides of such 3d metals as Fe, Co, Ni, Cu; it is also possible to use metallurgical dust. When using metal chlorides as metal containing phase, at the first stage water solutions of salts and polymers (PVA or PVC) are mixed. When mixing, the color changes in accordance with the complexes formed and when the water is removed xerogels in the form of colored films are formed. When at the first stage we apply metal oxides, the mechanic-chemical process is used in the presence of the active medium (water or water solution of acids or bases). Finally we also obtain the colored xerogels.

To investigate the processes at the initial stage, optical transmission microscopy, spectral photometry, IR and Raman spectroscopies, atomic force microscopy (AFM) are applied. For the corresponding correlations "polymer – metal containing phase" the dimensions, shape and energy characteristics of nanoreactors are found with the help of AFM [11, 12]. Depending on a metal participating in coordination, the structure and relief of xerogel surface change. The comparison of phase contrast pictures on the corresponding films indicates greater concentration of the extended polar structures in the films containing copper, in comparison with the films containing nickel and cobalt (Fig. 1.1). The processing of the pictures of phase contrast to reveal the regions of energy interaction of cantilever with the surface in comparison with the background produces practically similar result with optical transmission microscopy in Fig. 1.1.

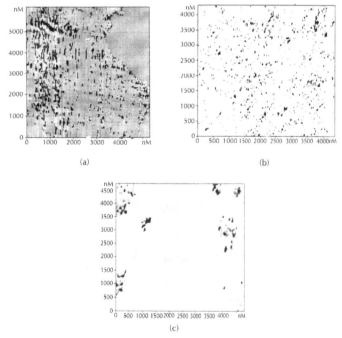

FIGURE 1.1   Pictures of phase contrast of PVA surfaces containing copper (a), nickel (b) and cobalt (c).

The results of AFM investigations of xerogels films (Fig. 1.2) obtained from metal oxides and PVA [15] is distinguished in comparison with previous data, that testify to difference in reactivity of metal chlorides and metal oxides.

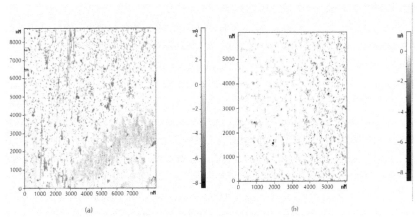

(a)   (b)

FIGURE 1.2   Phase contrast pictures of xerogels films PVA – Ni (a) and PVA – Cu (b).

Below (Fig. 1.3) the fields of energetic interaction of cantilever with surface of xerogels PVA – Ni (a) and PVA – Cu (b) are given.

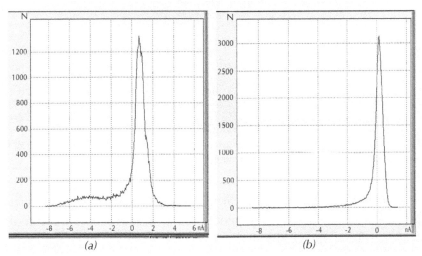

(a)   (b)

FIGURE 1.3   The pictures of energetic interaction of cantilever with surface of xerogels PVA – Ni (a) and PVA – Cu (b).

According to AFM results investigation, the addition of Ni/C nano-composite in PVA leads to more strong coordination in comparison with analogous addition Cu/C nanocomposite.

The mechanism of formation of nanoreactors filled with metals was found with the help of IR spectroscopy [11].

Thus, at the first stage, the coordination of metal containing phase and corresponding orientation in nanoreactor take place.

At the second stage, it is required to give the corresponding energy impulse to transfer the "transition state" formed into carbon/metal nanocomposite of definite size and shape. To define the temperature ranges in which the structuring takes place, DTA-TG investigation is applied. It is found that in the temperature range under 200°C nanofilms, from carbon fibers associated with metal phase as well, are formed on metal or metal oxide clusters. When the temperature elevates up to 400°C, 3D nanostructures are formed with different shapes depending on coordinating ability of the metal. In this case nanofilms are scrolled as seen in Fig. 1.4.

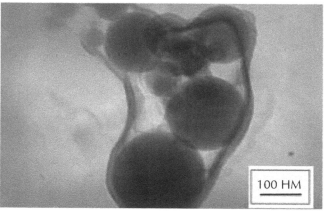

FIGURE 1.4   Microphotograph (transmission electron microscopy) demonstrating the moment of nanofilm scrolling on metal nanoparticles.

To investigate the processes at the second stage of obtaining metal/carbon nanocomposites X-ray photoelectron spectroscopy, transmission electron microscopy and IR spectroscopy are applied.

The sample for IR spectroscopy was prepared when mixing metal/carbon nanocomposite powder with 1 drop of vaselene oil in agate mortar to obtain a homogeneous paste with further investigation of the paste obtained on the appropriate instrument. As the vaselene oil was applied when the spectra were taken, we can expect strong bands in the range 2750–2950 cm^{-1}. Two types of nanocomposites rather widely applied during the modification of various polymeric materials were investigated. These were: copper/carbon nanocomposite and nickel/carbon nanocomposite specified below.

In turn, the nanopowders obtained were tested with the help of high-resolution transmission electron microscopy, electron microdiffraction, laser analyzer, X-ray photoelectron spectroscopy and IR spectroscopy.

The method of metal/carbon nanocomposite synthesis applied has the following advantages:

1. Originality of stage-by-stage obtaining of metal/carbon nanocomposites with intermediary evaluation of the influence of initial mixture composition on their properties.
2. Wide application of independent modern experimental and theoretical analysis methods to control the technological process.
3. Technology developed allows synthesizing a wide range of metal/carbon nanocomposites depending on the process conditions.
4. Process does not require the use of inert or reduction atmospheres and specially prepared catalysts.
5. Method of obtaining metal/carbon nanocomposites allows applying secondary raw material.

1.3.2 METHODS OF FINE DISPERSED SUSPENSIONS PREPARATION

To select the components of fine suspensions with the help of quantum-chemical modeling by the scheme described before [5], first, the interaction possibility of the material component being modified (or its solvent or surface-active substance) with metal/carbon nanocomposite is defined. The suspensions are prepared by the dispersion of the nanopowder in ultrasound station. The stability of fine suspension is controlled with the help of laser analyzer. The action on the corresponding regions participating in the formation of fine dispersed suspension or sol is determined with the help of IR spectroscopy. As an example, below one can see brief technique for obtaining fine suspension based on polyethylene polyamine.

Fine suspensions of metal/carbon nanocomposite were obtained mixing the nanopowder with polyethylene polyamine (PEPA) with further ultrasound processing on the stations Saphir UZV and UZTA-0.2/22-OM. The suspensions obtained were studied with the help of IR spectroscopy (IR-Fourier spectrophotometer FSM 1201).

1.4 RESULTS AND DISCUSSION

1.4.1 CHARACTERISTICS OF NANOCOMPOSITES OBTAINED

Under metal/carbon nanocomposite we understand the nanostructure containing metal clusters stabilized in carbon nanofilm structures. The carbon phase can be in the form of film structures or fibers. The metal particles are associated with carbon phase. The metal nanoparticles in the composite basically have the shapes close to spherical or cylindrical ones. Due to the stabilization and association of metal nanoparticles with carbon phase, chemically active metal particles are stable in air and during heating as the strong complex of metal nanoparticles with the matrix of carbon material is formed. The test results of nanocomposites obtained are given in Table 1.1.

TABLE 1.1 Characteristic of metal/carbon nanocomposites (Met/C HK).

Type Met/C HK	Cu/C	Ni/C	Co/C	Fe/C
Composition, (%): carbon (%)	50/50	60/40	65/35	70/30
Density, g/cm³	1.71	2.17	1.61	2.1
Average dimension, nm	20(25)	11	15	17
Specific surface, m²/g	160 (average)	251	209	168
Metal nanoparticle shape	Close to spherical, there are dodecahedrons	There are spheres and rods	Nanocrystals	Close to spherical

TABLE 1 *(Continued)*

Type Met/C HK	Cu/C	Ni/C	Co/C	Fe/C
Caron phase shape (shell)	Nanofibers associated with metal phase forming nanocoatings	Nanofilms scrolled in nanotubes	Nanofilms associated with nanocrystals of metal containing phase	Nanofilms forming nanobeads with metal containing phase
Atomic magnetic moment (reference), µB	0.0	0.6	1.7	2.2
Atomic magnetic moment (nanocomposite), µB	0.6	1.8	2.5	2.5

The nanocomposites described above were investigated with the help of IR spectroscopy by the technique indicated above. In this chapter, the IR spectra of Cu/C and Ni/C nanocomposites are discussed (Fig. 1.5), which find a wider application as the material modifiers.

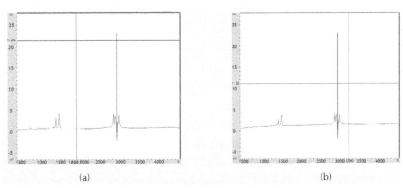

(a)   (b)

FIGURE 1.5   IR spectra of copper/carbon (a) and nickel/carbon (b) nanocomposite powder.

On IR spectra (Fig. 1.7) of two nanocomposites the common regions of IR radiation absorption are registered. Further, the bands appearing in the spectra and having the largest relative area were evaluated. We can see the difference in the intensity and number of absorption bands in the range 1300–1460 cm^{-1}, which confirms the different structures of composites.

In the range 600–800 cm^{-1} the bands with a very weak intensity are seen, which can be referred to the oscillations of double bonds (π-electrons) coordinated with metals. In case of Cu/C nanocomposite a weak absorption is found at 720 cm^{-1}. In case of Ni/C nanocomposite, except for this absorption, the absorption at 620 cm^{-1} is also observed.

In IR spectrum of copper/carbon nanocomposite two bands with a high relative area are found at:
- 1323 cm^{-1} (relative area – 9.28); and
- 1406 cm^{-1} (relative area – 25.18).

These bands can be referred to skeleton oscillations of polyarylene rings.

In IR spectrum of nickel/carbon nanocomposite the band mostly appears at 1406 cm^{-1} (relative area – 14.47).

According to the investigations with transmission electron microscopy the formation of carbon nanofilm structures consisting of carbon threads is characteristic for copper/carbon nanocomposite. In contrast, carbon fiber structures, including nanotubes, are formed in nickel/carbon nanocomposite. There are several absorption bands in the range 2800–3050 cm^{-1}, which are attributed to valence oscillations of C-H bonds in aromatic and aliphatic compounds. These absorption bonds are connected with the presence of vaselene oil in the sample. It is difficult to find the presence of metal in the composite as the metal is stabilized in carbon nanostructure. At the same time, it should be pointed out that apparently nanocomposites influence the structure of vaselene oil in different ways. The intensities and number of bands for Cu/C and Ni/C nanocomposites are different for:
- copper/carbon nanocomposite in the indicated range – 5 bands, and total intensity corresponds by the relative area – 64.63.
- nickel/carbon nanocomposite in the same range – 4 bands with total intensity (relative area) – 85.6.

The investigations of Carbon films in Metal/Carbon Nanocomposites and their peculiarities are carried out by Raman spectroscopy with the using of Laser Spectrometer Horiba LabRam HR 800. Below the Raman spectra and PEM microphotograph of Copper/Carbon (Cu/C) Nanocomposite are represented (Figs. 1.6 and 1.7).

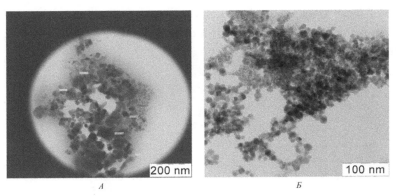

FIGURE 1.6 PEM microphotographs of Cu/C Nanocomposites (NC): A – Cu/C NC (28 nm); *B* – Cu/C NC (25 nm).

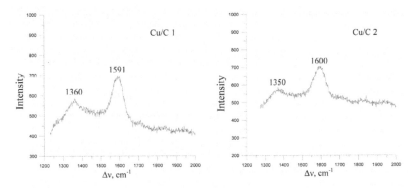

FIGURE 1.7 Raman spectra of Cu/C Nanocomposites: Cu/C 1 (28 nm); Cu/C 2 (25 nm).

Wave numbers and intensity relation testify to presence of nanoparticles containing the Copper atoms coordinated. At the same time the comparison of IR and Raman spectra shows their closeness on wave numbers and intensities relation.

1.4.2 CHARACTERISTICS OF METAL/CARBON NANOCOMPOSITES FINE DISPERSED SUSPENSIONS

The distribution of nanoparticles in water, alcohol and water-alcohol suspensions prepared based on the above technique are determined with the help of laser analyzer. In Fig. 1.8, one can see distributions of copper/carbon nanocomposite in the media different polarity and dielectric penetration.

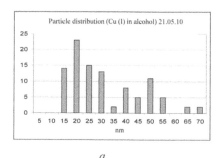 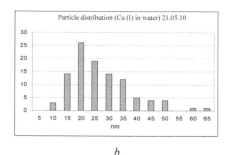

a   *b*

FIGURE 1.8   Distribution of copper/carbon nanocomposites in alcohol (a) in water (b).

When comparing the figures we can see that ultrasound dispersion of one and the same nanocomposite in media different by polarity results in the changes of distribution of its particles. In water solution the average size of Cu/C nanocomposite equals 20 nm, and in alcohol medium – greater by 5 nm.

Assuming that the nanocomposites obtained can be considered as oscillators transferring their oscillations onto the medium molecules, we can determine to what extent the IR spectrum of liquid medium will change, for example, polyethylene polyamine applied as a hardener in some polymeric compositions, when we introduce small and supersmall quantities of nanocomposite into it.

IR spectra demonstrate the change in the intensity at the introduction of metal/carbon nanocomposite in comparison with the pure medium (IR spectra are given in Fig. 1.9). The intensities of IR absorption bands are directly connected with the polarization of chemical bonds at the change in their length, valence angles at the deformation oscillations, that is, at the change in molecule normal coordinates.

When nanocomposites are introduced, the changes in the area and intensity of absorption bands, which indicates the coordination interactions and influence of nanostructure onto the medium are observed (Figs. 1.9–1.11).

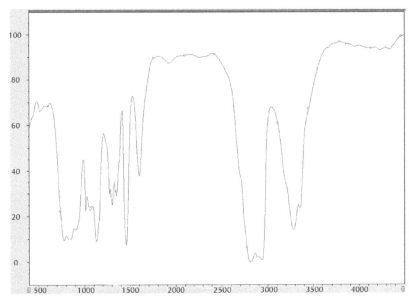

FIGURE 1.9   IR spectrum of polyethylene polyamine.

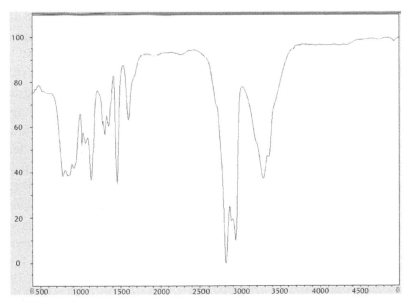

FIGURE 1.10   IR spectrum of copper/carbon nanocomposite fine suspension in polyethylene polyamine medium (ω (NC) = 1%).

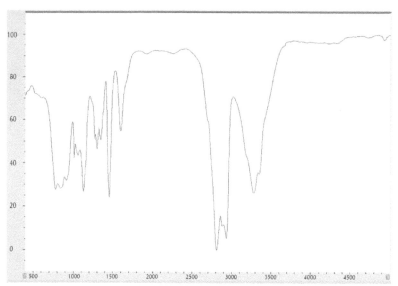

FIGURE 1.11 IR spectrum of nickel/carbon nanocomposite fine suspension in polyethylene polyamine (ω (NC) = 1%).

Special attention in PEPA spectrum should be paid to the peak at 1598 cm^{-1} attributed to deformation oscillations of N-H bond, where hydrogen can participate in different coordination and exchange reactions.

In the spectra wave numbers characteristic for symmetric $v_s(NH_2)$ 3352 cm^{-1} and asymmetric $v_{as}(NH_2)$ 3280 cm^{-1} oscillations of amine groups are present. There is a number of wave numbers attributed to symmetric $v_s(CH_2)$ 2933 cm^{-1} and asymmetric valence $v_{as}(CH_2)$ 2803 cm^{-1}, deformation wagging oscillations $v_D(CH_2)$ 1349 cm^{-1} of methylene groups, deformation oscillations of NH v_D (NH) 1596 cm^{-1} and NH$_2$ $v_D(NH_2)$ 1456 cm^{-1} amine groups. The oscillations of skeleton bonds at $v(CN)$ 1059–1273 cm^{-1} and $v(CC)$ 837 cm-^{1} are the most vivid. The analysis of intensities of IR spectra of PEPA and fine suspensions of metal/carbon nanocomposites based on it revealed a significant change in the intensities of amine groups of dispersion medium (for $v_s(NH_2)$ in 1.26 times, and for $v_{as}(NH_2)$ in approximately 50 times).

Such demonstrations are presumably connected with the distribution of the influence of nanoparticle oscillations onto the medium with further structuring and stabilizing of the system. Under the influence of nanoparticle the medium changes which is confirmed by the results of IR spectroscopy (change in the intensity of absorption bands in IR region). Densities, dielectric penetration, viscosity of the medium are the determining parameters for obtaining fine suspension with uniform distribution of particles in the volume. At the same time, the structuring rate and consequently the stabilization of the system directly depend on the distribution by particle sizes in suspension. At the wide range of particle distribution by sizes, the oscillation frequency of particles different by size can significantly differ, in this connection, the distortion in the influence transfer of nanoparticle system onto the medium is possible (change in the medium from the part of some particles can be balanced by the other). At the narrow range of nanoparticle distribution by sizes the system structuring and stabilization are possible. With further adjustment of the components such processes will positively influence the processes of structuring and self-organization of final composite system determining physical-mechanical characteristics of hardened or hard composite system.

The effects of the influence of nanostructures at their interaction into liquid medium depend on the type of nanostructures, their content in the medium and medium nature. Depending on the material modified, fine suspensions of nanostructures based on different media are used. Water and water solutions of surface-active substances, plasticizers, foaming agents (when modifying foam concretes) are applied as such media to modify silicate, gypsum, cement and concrete compositions. To modify epoxy compounds and glues based on epoxy resins the media based on polyethylene polyamine, isomethyltetrahydrophthalic anhydride, toluene and alcohol-acetone solutions are applied. To modify polycarbonates and derivatives of polymethyl methacrylate dichloroethane and dichloromethane media are used. To modify polyvinyl chloride compositions and compositions based on phenolformaldehyde and phenolrubber polymers alcohol or acetone-based media are applied. Fine suspensions of metal/carbon nanocomposites are produced using the above media for specific compositions. In IR spectra of all studied suspensions the significant change in the absorption intensity, especially in the regions of wave numbers close to the corresponding nanocomposite oscillations, is observed. At the same time, it is found that the effects of nanocomposite influence on liquid media (fine suspensions) decreases with time and the activity of the corresponding suspensions drops. The time period in which the appropriate activity of nanocomposites is kept changes in the interval 24 h – 1 month depending on the nanocomposite type and nature of the basic medium (liquid phase in which nanocomposites dispergate). For instance, IR spectroscopic investigation of fine suspension based on isomethyltetrahydrophthalic anhydride containing 0.001% of Cu/C nanocomposite indicates the decrease in the peak intensity, which sharply increased on the third day when nanocomposite was introduced (Fig. 1.12).

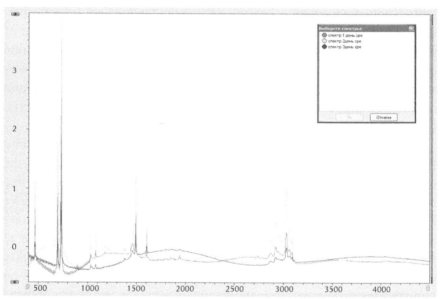

FIGURE 1.12   Changes in IR spectrum of copper/carbon nanocomposite fine suspension based on isomethyltetrahydrophthalic anhydride with time. (a – IR spectrum on the first day after the nanocomposite was introduced, *B* – IR spectrum on the second day, c – IR spectrum on the third day).

Similar changes in IR spectra take place in water suspensions of metal/carbon nanocomposites based on water solutions of surface-active nanocomposites. In Fig. 1.13, one can see IR spectrum of iron/carbon nanocomposite based on water solution of sodium lignosulfonate in comparison with IR spectrum of water solution of surface-active substance.

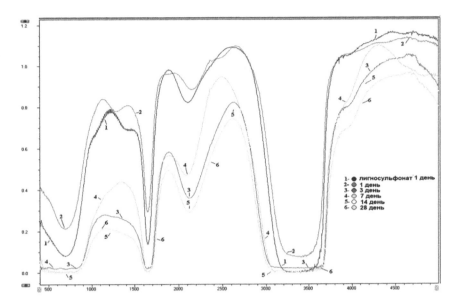

FIGURE 1.13 Comparison of IR spectra of water solution of sodium lignosulfonate (1) and fine suspension of iron/carbon nanocomposite (0.001%) based on this solution on the first day after nanocomposite introduction (2), on the third day (3), on the seventh day (4), 14th day (5) and 28th day (6).

As it is seen, when nanocomposite is introduced and undergoes ultrasound dispergation, the band intensity in the spectrum increases significantly. Also the shift of the bands in the regions 1100–1300 cm-¹, 2100–2200 cm-¹ is observed, which can indicate the interaction between sodium lignosulfonate and nanocomposite. However after two weeks the decrease in band intensity is seen. As the suspension stability evaluated by the optic density is 30 days, the nanocomposite activity is quite high in the period when IR spectra are taken. It can be expected that the effect of foam concrete modification with such suspension will be revealed if only 0.001% of nanocomposite is introduced.

The ultimate breaking stresses were compared in the process of compression of foam concretes modified with copper/carbon nanocomposites obtained in different nanoreactors of polyvinyl alcohol [13, 14]. The sizes of nanoreactors change depending on the crystallinity and correlation of acetate and hydroxyl groups in PVA, which results in the change of sizes and activity of nanocomposites obtained in nanoreactors. It is observed that the sizes of nanocomposites obtained in nanoreactors of PVA matrixes 16/1(ros) (NC2), PVA 16/1 (imp) (NC1), PVA 98/10 (NC3), correlate as NC3 > NC2 > NC1. The smaller the nanoparticle size and greater its activity, the less amount of nanostructures is required for self-organization effect. At the same time, the oscillatory nature of the influence of these nanocomposites on the compositions of foam concretes is seen in the fact that if the amount of nanocomposite is 0.0018% from the cement mass, the significant decrease in the strength of NC1 and NC2 is observed. The increase in foam concrete strength after the modification with iron/carbon nanocomposite is a little smaller in comparison with the effects after the application of NC1 and NC2 as modifiers. The corresponding effects after the modification of cement, silicate, gypsum and concrete compositions with nanostructures is defined by the features of components and technologies applied. These features often explain the instability of the results after the modification of the foregoing compositions with nanostructures. Besides during the modification the changes in the activity of fine suspensions of nanostructures depending on the duration and storage conditions should be taken into account.

In this regard, it is advisable to use metal/carbon nanocomposites when modifying polymeric materials whose technology was checked on strictly controlled components.

At present a wide range of polymeric substances and materials: compounds, glues, binders for glass-, basalt and carbonplastics based on epoxy resins, phenol-rubber compositions, polyimide and polyimide compositions, materials on polycarbonate and polyvinyl chloride basis, as well as special materials, such as current-conducting glues and pastes, fireproof intumescent glues and coatings are being modified.

Below you can find the results of some working-outs [15]:

1.   The introduction of metal/carbon nanostructures (0.005%) in the form of fine suspension into polyethylene polyamine or the mixture of amines into epoxy compositions allows increasing the thermal stability of the compositions by 75–100 degrees and consequently increase the application range of the existing products. This modification contributes to the increase in adhesive and cohesive characteristics of glues, lacquers and binders.

2. Hot vulcanization glue was modified with copper/carbon and nickel/carbon nanostructures using toluol-based fine suspensions. On the test results of samples of four different schemes the tear strength σ_t increased up to 50% and shear strength τ_s – up to 80%, concentration of metal/carbon nanocomposite introduced was 0.0001–0.0003%.

3. Fine suspension of nanostructures was produced in polycarbonate and dichloroethane solutions to modify polycarbonate-based compositions. The introduction of 0.01% of copper/carbon nanostructures leads to the significant decrease in temperature conductivity of the material (in 1.5 times). The increase in the transmission of visible light in the range 400–500 nm and decrease in the transmission in the range 560–760 nm were observed.

4. When modifying polyvinyl chloride film with fine suspension containing iron/carbon nanocomposite, the increase of the crystalline phase in the material was observed. The PVC film modified containing 0.0008% of NC does not accumulate the electrostatic charge on its surface. The material obtained completely satisfies the requirements applied to PVC films for stretch ceilings.

5. The introduction of nickel/carbon nanocomposite (0.01% of the mass of polymer filled on 65% of silver microparticles) into the epoxy polymer hardened with polyethylene polyamine leads to the decrease in electric resistance to 10^{-5} Ohm×cm (10^{-4} Ohm×cm without nanocomposite).

1.5 CONCLUSION

In this chapter, the possibilities of developing new ideas about self-organization processes and about nanostructures and nanosystems are discussed on the example of metal/carbon nanocomposites. It is proposed to consider the obtaining of metal/carbon nanocomposites in nanoreactors of polymeric matrixes as self-organization process similar to the formation of ordered phases, which can be described with Avrami equation. The application of Avrami equations during the synthesis of nanofilm structures containing copper clusters has been tested. The influence of nanostructures on active media is given as the transfer of oscillation energy of the corresponding nanostructures onto the medium molecules.

IR spectra of metal/carbon and their fine suspensions in different (water and organic) media have been studied for the first time. It has been found that the introduction of supersmall quantities of prepared nanocomposites leads to the significant change in band intensity in IR spectra of the media. The attenuation of oscillations generated by the introduction of nanocomposites after the time interval specific for the pair "nanocomposite – medium" has been registered.

Thus to modify compositions with fine suspensions it is necessary for the latter to be active enough that should be controlled with IR spectroscopy.

A number of results of material modification with fine suspensions of metal/carbon nanocomposites are given, as well as the examples of changes in the properties of modified materials based on concrete compositions, epoxy and phenol resins, polyvinyl chloride, polycarbonate and current-conducting polymeric materials.

KEYWORDS

- **metal/carbon nanocomposites**
- **modification**
- **nanochemistry methods**
- **super small quantities of nanocomposites**
- **synthesis in nanoreactors**

REFERENCES

1. Kodolov V. I., Khokhriakov N. V. Chemical physics of the processes of formation and transformation of nanostructures and nanosystems. – Izhevsk: Izhevsk State Agricultural Academy, 2009. In 2 volumes. V. 1. 360p. V. 2. 415p.
2. Melikhov I. V., Bozhevolnov V. E. Variability and self-organization in nanosystems. J. Nanoparticle Research, 2003.V. 5. P. 465–472.

3. Malinetsky G. G. Designing of the future and modernization in Russia. Preprint of M. V. Keldysh Institute of Applied Mechanics, 2010. Issue 41. 32p.
4. Tretyakov Yu.D. Self-organization processes. Uspekhi Chimii, 2003. V.72. Issue 8. P. 731–764.
5. Kodolov V. I., Khokhriakov N. V., Trineeva V. V., Blagodatskikh I. I. Activity of nanostructures and its representation in nanoreactors of polymeric matrixes and active media. Chemical physics and mesoscopy, 2008. V.10. Issue 4. P. 448–460.
6. Wunderlich B. Physics of macromolecules. In 3 volumes. M.: Mir, 1979. V. 2. 574p.
7. Fedorov V. B., Khakimova D. K., Shipkov N. N., Avdeenko M.A. To thermodynamics of carbon materials. Doklady AS USSR, 1974. V. 219. Issue 3. P. 596–599; Fedorov V. B., Khakimova D. K., Sharshorov M. H. et al. To kinetics of graphitation. Doklady AS USSR, 1975. V. 222. Issue 2. p. 399–402.
8. Theory of chemical fibers formation. ed. by Serkov A.T. – M.: Chemistry, 1975. 548p.
9. Kodolov V. I., Khokhriakov N. V., Kuznetsov A. P. To the issue of the mechanism of the influence of nanostructures on structurally changing media at the formation of "intellectual" composites. Nanotechnics, 2006. 3(7). p. 27–35.
10. Kodolov V. I., Khokhriakov N. V., Trineeva V. V., Blagodatskikh I. I. Problems of Nanostructure Activity Estimation, Nanostructures Directed Production and Application. Nanomaterials Yearbook – 2009. From nanostructures, nanomaterials and nanotechnologies to Nanoindustry – N. Y.: Nova Science Publishers, Inc., 2010. p. 1–18.
11. Kodolov V. I., Blagodatskikh I. I., Lyakhovich A. M. et al. Investigation of the formation processes of metal containing carbon nanostructures in nanoreactors of polyvinyl alcohol at early stages. Chemical physics and mesoscopy, 2007. V. 9. Issue 4. p. 422–429.
12. Trineeva V. V., Lyakhovich A.M., Kodolov V. I. Forecasting of the formation processes of carbon metal containing nanostructures using the method of atomic force microscopy. Nanotechnics, 2009. 4 (20). p. 87–90.
13. Akhmetshina L. F., Kodolov V. I., Tereshkin I. P., Korotin A. I. Influence of carbon metal containing nanostructures on strength of concrete composites. Nanotechnologies in construction, 2010. Issue 6. p. 35–46.
14. Kodolov V. I., Trineeva V. V., Kovyazina O. A., Vasil'chenko Yu.M. Production and application of metal/carbon nanocomposites. p. 27–35.
15. Kovyazina O.A., Trineeva V. V., Akhmetshina L. F., Vasilchenko Yu.M. et al. Experience of the application of metal/carbon nanocomposites to modify materials. Abstracts of VII International scientific-practical conference "Nanotechnologies for production-2010." Fryazino, p. 53–54.

CHAPTER 2

THE METAL/CARBON NANOCOMPOSITES SYNTHESIS IN NANOREACTORS OF POLYMERIC MATRIXES

V. I. KODOLOV, V. V. TRINEEVA, YU. M. VASIL'CHENKO, and YA. A. POLYOTOV

CONTENTS

ABSTRACT

The Redox synthesis of Metal/Carbon Nanocomposites in the nanoreactors of polymeric matrixes is proposed. Two methods of Metal/Carbon Nanocomposites are observed. The first method is comprised in the gel-like colored films obtaining from solutions mixture of metal chlorides and polyvinyl alcohol. The films are heated according to the program. The nanoproduct obtained is dispersed in corresponding medium by ultrasound dispergator. The synthesis mechanism in polymeric matrixes nanoreactors is considered. The essence of the second method consists in coordination of functional groups of polymer and compounds of 3d-metals as a result of grinding of metal containing and polymer phases. For this case mechanism of nanocomposite formation in nanoreactors of polymeric matrixes is represented, and also nanoreactors designing problems are discussed. The Redox synthesis conditions and the Metal/Carbon Nanocomposites characteristics are given.

2.1   INTRODUCTION

The perspective area of nanochemistry is the chemistry in nanoreactors. Nanoreactors can be compared with specific nanostructures representing limited space regions in which chemical particles orientate creating "transition state" prior to the formation of the desired nanoproduct. Nanorectors have a definite activity, which predetermines the creation of the corresponding product. When nanosized particles are formed in nanoreactors, their shape and dimensions can be the reflection of shape and dimensions of the nanoreactor.

The proposed method of corresponding nanostructures synthesis consists in the conducting of redox processes, which proceed in nanoreactors of polymeric matrixes, and is accompanied by the reduction of metal ions included into the cavities of organic polymer gels. At the same time, hydrocarbon shells are simultaneously oxidized to carbon.

2.2 THE SYNTHESIS OF METAL/CARBON NANOCOMPOSITES IN NANOREACTORS OF POLYMERIC MATRIXES FROM SOLUTIONS OF POLYMERS AND METAL SALTS

The process starts with the preparation of polymer solutions, for instance, polyvinyl alcohol (PVA), and metal compounds, for instance, 3d-metal chlorides. Afterwards, the solutions with a certain concentration are mixed in the ratio "PVA-metal chloride" equals to 20:1–1:5 (better 5:1). Then the prepared solutions are dried till they obtain gel-like colored films with further temperature elevation up to 100°C. The films obtained are controlled by spectral photometry, and also with help of transmission optical microscopy, atomic force microscopy and X-ray photoelectron spectroscopy. When the film color changes to black, the films are heated in the furnace according to the following program: 100–200–300–400°C. As a result, the dark porous semiproduct with many microcracks is formed, that is milled in spherical or jet mill. The nanopowder obtained is steamed and dispersed in hot water. After filtration, the powder is dried and tested with the help of Raman spectroscopy, X-ray photoelectron spectroscopy, transmission electron microscopy and electron microdiffraction.

When PVA was added to the powders of metal chlorides, the color of the mixture changed: the mixture of copper chloride became yellow-green, the cobalt chloride mixture – blue, and nickel chloride – pale-green (Fig. 2.1).

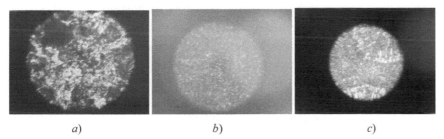

a) b) c)

FIGURE 2.1 The photographs of the samples containing PVA and copper chloride (a), cobalt chloride (b), and nickel chloride (c).

Observing the color changes, one can draw a conclusion that when PVA interacts with metal chlorides, the formation of complex compounds takes place.

Among the above-discussed metals, iron is most active. Brown-red inclusions on the photograph evidence the formation of the complex iron compounds. In addition, on all the photographs depicting the mixtures containing metal chlorides, one can see a net of weaves, which are most likely the reflections of nanostructures.

In order to compare these structures, the investigations of the morphology of the films changing over a certain range of temperatures were carried out with the help of atomic power microscopy. When the nanoproduct pictures obtained by atomic power microscopy and optical microscopy are compared with the TEM micrograph of the nanoproduct treated thermally and with aqueous solution for the matrix removal, one can notice some correspondence between them. The nanoproduct represents interweaving tubulens containing Cu(I), Cu(II). In Fig. 2.1, there are also optical effects indicating light polarization at light transmission through the films owing to the defects appearing during the formation of the complex compounds at the initial stage of the process.

Due to the fact, that metal ions are active, in the polymer medium they immediately appear in the environment of the PVA molecules and form bonds with the hydroxyl groups of this polymer. Polyvinyl alcohol replicates the structure of the particle that it surrounds; however, due to the tendency of the molecules of the metal salts or other metal compounds to combine, PVA as if envelopes the powder particles, and therefore the forms of the obtained nanostructures can be different. The optical microscopy method allows to determine the structure of the nanocomposite at the early stage.

When the samples are heated, dehydration occurs, and as a result, metal-containing nanotubes form. These processes are thoroughly described in works [1, 2]. Dehydration leads to the darkening of the film. After the samples have been heated, on the photograph the remaining net of weaves can be seen, that is, the structure morphology has remained. To some extent, this fact indicates that the initially formed structure of matrices is inherited. The methods of optical spectroscopy and X-ray photoelectron spectroscopy allow to determine the energy of the interaction of the chemical particles in the nanoreactors with the active centers of the nanoreactor walls, which stimulate reduction-oxidation processes.

Depending on the nature of the metal salt and the electrochemical potential of the metal, different metal reduction nanoproducts in the carbon shells differing in shape are formed. Based on this result we may speak about a new scientific branch – nanometallurgy. The stages of nanostructures synthesis may be represented by the following scheme (Fig. 2.2).

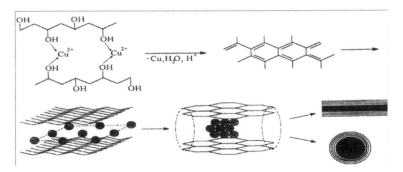

FIGURE 2.2   Scheme of copper-carbon nanostructures obtaining from copper ions and polyvinyl alcohol.

The possible ways for obtaining metallic nanostructures in carbon shells have been determined. The investigation results allow to speak about the possibility of the isolation of metallic and metal-containing nanoparticles in the carbon shells differing in shape and structure. However, there are still problems related to the calculation and experiment because using the existing investigation methods it is difficult unambiguously to estimate the geometry and energy parameters of nanoreactors under the condition of 'erosion' of their walls during the formation of metallic nanostructures in them.

2.3   METHODS OF METAL/CARBON NANOCOMPOSITES PRODUCTION IN POLYMERIC MATRIX NANOREACTORS FILLED BY METAL CONTAINING PHASES

The metal/carbon nanocomposites production proceeds in two stages. At the first stage, the nanoreactor, filled by the corresponding metal containing phase, is formed in the determined polymeric matrix (polyvinyl alcohol, polyvinyl chloride, polyvinyl acetate). At the second stage, metal/carbon nanocomposite is formed with the metal containing phase reduction and simultaneously the polymeric phase hydrocarbon part oxidation.

Because of the association of polymeric phase with metal containing phase the walls of nanoreactor participate in redox process with cover formation for metal containing clusters obtained [3–6]. Metal/carbon nanocomposites will be more active than carbon or silicon nanostructures because their masses are bigger at identical sizes and shapes. Therefore the vibration energy transmitted to the medium is also high.

Metal/carbon nanocomposite (Me/C) represents metal nanoparticles stabilized in carbon nanofilm structures [7–9]. In turn, nanofilm structures are formed with carbon amorphous nanofibers associated with metal containing phase. As a result of stabilization and association of metal nanoparticles with carbon phase, the metal chemically active particles are stable in the air and during heating as the strong complex of metal nanoparticles with carbon material matrix is formed. The test results of nanocomposites obtained are given in Table 2.1.

TABLE 2.1   Characteristics of Metal/Carbon nanocomposites (Met/C NC).

Type of Met/C NC	Cu/C	Ni/C	Co/C	Fe/C
Composition, Metal/Carbon [%]	50/50	60/40	65/35	70/30
Density, [g/cm^3]	1.71	2.17	1.61	2.1
Average dimension, [nm]	20(25)	11	15	17
Specific surface, [m^2/g]	160 (average)	251	209	168
Metal nanoparticle shape	Close to spherical, there are dodecahedrons	There are spheres and rods	Nanocrystals	Close to spherical
Caron phase shape (shell)	Nanofibers associated with metal phase forming nanocoatings	Nanofilms scrolled in nanotubes	Nanofilms associated with nanocrystals of metal containing phase	Nanofilms forming nanobeads with metal containing phase
Atomic magnetic moment [8], [μB]	0.0	0.6	1.7	2.2
Atomic magnetic moment (nanocomposite), [μB]	0.6	1.8	2.5	2.5

For the corresponding correlations "polymer – metal containing phase" the dimensions, shape and energy characteristics of nanoreactors are found with the help of AFM [4, 10]. Depending on a metal participating in coordination, the structure and relief of xerogel surface change.

To investigate the processes at the second stage of obtaining metal/carbon nanocomposites X-ray photoelectron spectroscopy, transmission electron microscopy and IR spectroscopy are applied.

In turn, the nanopowders obtained were tested with the help of high-resolution transmission electron microscopy, electron microdiffraction, laser analyzer, X-ray photoelectron spectroscopy and IR spectroscopy.

The distinctive feature of the considered technique for producing metal/carbon nanocomposites is a wide application of independent, modern, experimental and theoretical analysis methods to substantiate the proposed technique and investigation of the composites obtained (quantum-chemical calculations, methods of transmission electron microscopy and electron diffraction, method of X-ray photoelectron spectroscopy, X-ray phase analysis, etc.). The technique developed allows synthesizing a wide range of metal/carbon nanocomposites by composition, size and morphology depending on the process conditions. In its application it is possible to use secondary metallurgical and polymer raw materials. Thus, we can adjust the nanocomposite structure to extend the function of its application without prefunctionalization. Controlling the sizes and shapes of nanostructures by changing the metal-containing phase, we can, to some extent, apply completely new, practicable properties to the materials which sufficiently differ from conventional materials.

The essence of the method [6] consists in coordination interaction of functional groups of polymer and compounds of 3d-metals as a result of grinding of metal-containing and polymer phases. Further, the composition obtained undergoes thermolysis following the temperature mode set with the help of thermo gravimetric and differential thermal analyzes. At the same time, we observe the polymer carbonization, partial or complete reduction of metal compounds and structuring of carbon material in the form of nanostructures with different shapes and sizes.

Metal/carbon nanocomposite (Me/C) represents metal nanoparticles stabilized in carbon nanofilm structures [7, 8]. In turn, nanofilm structures are formed with carbon amorphous nanofibers associated with metal containing phase. As a result of stabilization and association of metal nanoparticles with carbon phase, the metal chemically active particles are stable in the air and during heating as the strong complex of metal nanoparticles with carbon material matrix is formed.

Figure 2.3 demonstrates the microphotographs of transmission electron microscopy specific for different types of metal/carbon nanocomposites.

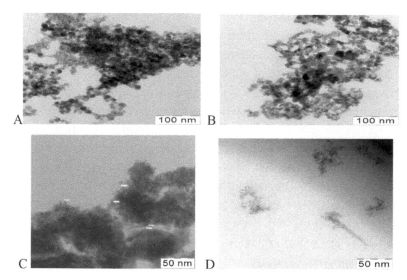

FIGURE 2.3 Microphotographs of metal/carbon nanocomposites: A – Cu/C; *B* – Ni/C; C – Co/C; D – Fe/C.

One of the main properties of metal/carbon nanocomposites obtained is the ability to form fine suspensions [9] in various media (organic solvents, water, solutions of surface-active substances). The average size of nanoparticles in fine suspensions is 10–25 nm depending on the type of metal/carbon nanocomposite.

The short information about nanostructures formation mechanism in polymeric matrix nanoreactors as well as about the methods of synthesis and control during metal/carbon nanocomposites production represents. The main attention is given for the ability of nanocomposites obtained to form the fine dispersed suspensions in different media and for the distribution of nanoparticles in media. The examples of improving technical characteristics of foam concretes and glue compositions are given.

2.4 NANOREACTORS DESIGNING PROBLEMS AND THE METHODS OF FORMATION OF NANOREACTORS FILLED BY METAL CONTAINING PHASE

For the corresponding correlations "polymer – metal containing phase" the dimensions, shape and energy characteristics of nanoreactors are found with the help of AFM [4, 10]. Depending on a metal participating in coordination, the structure and relief of xerogel surface change. The comparison of phase contrast pictures on the corresponding films indicates greater concentration of the extended polar structures in the films containing copper, in comparison with the films containing nickel and cobalt. The processing of the pictures of phase contrast to reveal the regions of energy interaction of cantilever with the surface in comparison with the background produces practically similar result with optical transmission microscopy. Corresponding to data of AFM the sizes of nanoreactors obtained from solutions of metal chlorides and the mixture of polyvinyl alcohol (PVA) with polyethylene polyamine (PEPA) are determined (Table 2.2).

TABLE 2.2 Sizes of nanoreactors found with the help of Atomic Force Microscopy.

Composition	Sizes of AFM formations				
	Length	Width	Height	Area	Density
PVA:PEPA: CoCl2= 2:1:1	400–800	150–400	30–40	60–350	5.5
PVA: PEPA: NiCl2=2:1:1	80–100	80–100	25–35	6–12	120
PVA: PEPA: CuCl2=2:1:1	80–100	80–100	20–30	6–20	20
PVA: PEPA: CoCl2=2:2:1	600–900	300–600	100–120	180–500	3.0
PVA: PEPA: NiCl2=2:2:1	40–80	40–60	10–30	2–4	350

The results of AFM investigations of xerogels films obtained from metal oxides and PVA [10] is distinguished in comparison with previous data that testify to difference in reactivity of metal chlorides and metal oxides.

According to AFM results investigation the addition of Ni/C nanocomposite in PVA leads to more strong coordination in comparison with analogous addition Cu/C nanocomposite.

The mechanism of formation of nanoreactors filled with metals was found with the help of IR spectroscopy.

Depending on metal coordinating ability and conditions for nanostructure obtaining (in liquid or solid medium with minimal content of liquid) we obtain "embryos" of future nanostructures of different shapes, dimensions and composition. It is advisable to model coordination processes and further redox processes with the help of quantum chemistry apparatus.

The availability of d metal in polymeric matrix results, in accordance with modeling results, in its regular distribution in the matrix and self-organization of the matrix.

At the same time, the metal orientation proceeds in interface regions and nanopores of polymeric phase which conditions further direction of the process to the formation of metal/carbon nanocomposite. In other words, the birth and growth of nanosize structures occur during the process in the same way as known from the macromolecule physics [11].

2.5   THE CONDITIONS OF REDOX SYNTHESIS OF METAL/ CARBON NANOCOMPOSITES AND THE NANOCOMPOSITES CHARACTERIZATION

To define the temperature ranges in which the structuring takes place, DTA-TG investigation is applied. It is known that small changes of weight loss (TG curve) at invariable exothermal effect (DTA curve) testify to the self-organization (structural formation) in system.

According to data of DTA-TG investigation optimal temperature field for film nanostructure obtaining is 230–270°C, and for spatial nanostructure obtaining – 325–410°C.

It is found that in the temperature range under 200°C nanofilms, from carbon fibers associated with metal phase as well, are formed on metal or metal oxide clusters. When the temperature elevates up to 400°C, 3D nanostructures are formed with different shapes depending on coordinating ability of the metal.

Two types of nanocomposites (in vaselene oil) were investigated by IR spectroscopy.

In case of Cu/C nanocomposite a weak absorption is found at 720 cm^{-1}. In case of Ni/C nanocomposite, except for this absorption, the absorption at 620 cm^{-1} is also observed.

In IR spectrum of copper/carbon nanocomposite two bands with a high relative area are found at:
- 1323 cm^{-1} (relative area – 9.28)
- 1406 cm^{-1} (relative area – 25.18).

These bands can be referred to skeleton oscillations of polyarylene rings.

In IR, spectrum of nickel/carbon nanocomposite the band mostly appears at 1406 cm^{-1} (relative area – 14.47).

According to the investigations with transmission electron microscopy the formation of carbon nanofilm structures consisting of carbon threads is characteristic for copper/carbon nanocomposite. In contrast, carbon fiber structures, including nanotubes, are formed in nickel/carbon nanocomposite.

Assuming that the nanocomposites obtained can be considered as oscillators transferring their oscillations onto the medium molecules, we can determine to what extent the IR spectrum of liquid medium will change. It is proposed to consider the obtaining of metal/carbon nanocomposites in nanoreactors of polymeric matrixes as self-organization process similar to the formation of ordered phases.

The perspectives of this investigation are looked through in an opportunity of thin regulation of processes and the entering of corrective amendments during processes.

2.6   CONCLUSION

Among the many various methods of obtaining metal clusters and carbon/ metal containing nanostructures the synthesis in nanoreactors is considered the most perspective [3, 12]. The main advantage of the above synthetic method in comparison with the others is the possibility to predict the result obtained depending on the matrix and nanoreactor selected, as well as the basic metal. However this method has not been sufficiently developed. For its development theoretical and experimental results are required. This paper discusses the ways for the production and investigation of nanoreactors in polymeric matrixes and synthesis of nanostructures with corresponding activity in them. Nanocomposites obtained in nanoreactors can be applied as modifiers of different compositions and materials.

KEYWORDS

- **3D metals compounds**
- **grinding**
- **Metal/Carbon nanocomposites**
- **mixtures**
- **nanoreactors, polymeric matrixes**
- **program of heat**
- **redox synthesis**
- **solutions**
- **ultrasound dispersion**

REFERENCES

1. Volkova I. E., Volkov A.Yu., Murzakaev A. M. et al. The Physics of Metals and Metallography, 2003 v.95, No 4, p.342–345
2. Didik A. A., Kodolov V. I., Volkov A.Yu. et al. Inorg.Mat., 2003, v.39, No 6, p.693–697
3. Kodolov V. I., Khokhriakov N. V. Chemical physics of the processes of nanostructure and nanosystem formation and transformation. – Izhevsk: Publishing House of Izhevsk State Agricultural Academy, 2009. V. 1.–361p. V. 2.–415p

4. Kodolov V. I., Khokhriakov N. V., Trineeva V. V., and Blagodatskikh I. I. Activity of nanostructures and its presence in nanoreactors of polymeric matrixes and active media. Chem. Phys. Mesoscopy, 2008. V. 10. *n* 4. Pp. 448–460.
5. Kodolov V. I. Commentary to the paper by V. I. Kodolov et al. Chem. Phys. Mesoscopy, 2009. V. 11. *n* 1. Pp. 134–136.
6. Vasil'chenko Yu.M., Akhmetshina L. F., Shklyeva D. A. et al. Synthesis of carbon metal containing nanostructures in PVC and PVAc gels with metallurgical dust. in Nanomaterials Yearbook – 2009. From nanostructures, nanomaterials and nanotechnologies to nanoindustry. – N. Y.: Nova Sci. Publ. Inc., 2010. – Pp. 283–288.
7. Kodolov V. I., Kodolova V. V., Semakina N. V. et al. Patent RU 2337062. 27.10.2008.
8. Kodolov V. I., Vasil'chenko Yu.M., Shklyeva D. A. et al. Patent RU 2393110. 27.06.2010.
9. Kodolov V. I., Vasil'chenko Yu.M., Akhmetshina L. F. et al. Patent RU 2423317. 10.07.2011
10. Trineeva V. V., Lyakhovitch A. M., Kodolov V. I. Prognosis of metal containing carbon nanostructures formation processes with the using of AFM method. Nanotechnics, 2009. Issue 20. Pp. 87–90.
11. Vunderlikh B. Physics of macromolecules. V. 2. – M.: Mir, 1979. 574p.
12. Bronstein L. M. and Z. B. Shifrina. Nanoparticles in dendrimers: from synthesis to application. Russian nanotechnologies, 2009. Vol. 4. Issue 9–10. pp. 32–54.

CHAPTER 3

FUNDAMENTALS OF PRODUCTION TECHNOLOGY OF CARBON-METAL-CONTAINING NANOPRODUCTS IN NANOREACTORS OF POLYMERIC MATRIXES

V. I. KODOLOV, V. V. TRINEEVA, O. A. KOVYAZINA, and A. YU. BONDAR

CONTENTS

ABSTRACT

The technology proposed consists in the conducting of redox processes, which proceed in nanoreactors of polymeric matrixes and are accompanied by the reduction of metal ions included into the cavities of organic polymer gels. At the same time, hydrocarbon shells are simultaneously oxidized to carbon.

The process starts with the preparation of polymer solutions, for instance, polyvinyl alcohol (PVA), and metal compounds, for instance, 3d-metal chlorides. Afterwards, the solutions with a certain concentration are mixed in the ratio "PVA-metal chloride" equals to 20:1–1:5. Then the prepared solutions are dried till they obtain gel-like colored films with further temperature elevation up to 100°C. The color changes depending upon the metal. The films obtained are controlled spectrophotometrically, and also with the help of transmission optical microscopy and atomic force microscopy. When the film color changes to black, the films are heated in the furnace according to the following program: 100–200–300–400°C. As a result, the dark porous semiproduct with many microcracks is formed, that is milled in spherical or jet mill. The nanopowder obtained is steamed and dispersed in hot water. After filtration, the powder is dried and tested by means of Raman spectroscopy, X-ray photoelectron spectroscopy, transmission electron microscopy and electron microdiffraction.

As a result, carbon-metal-containing multiwall nanotubes and astralenes are obtained. Depending upon the conditions nanoproducts can be metallic nanoparticles and nanowires in carbon shells. The yield of nanoproducts, calculated on carbon of PVA, is about 80–90%. The nanoproducts obtained can be applied as additives (0.01–0.05%) to improve the properties of foam concrete and intumescent fire-resistant coatings whose characteristics improve in 2–3 times.

3.1   INTRODUCTION

Chemistry in nanoreactors has been rapidly developed during the last few years. At present, the term "nanoreactor" is regarded in a wider implication. As for the modern definition, the term "nanoreactor" may include the defective regions of metal salt polycrystals, interface regions in substances with lamellar structures, extended cavities formed by macromolecules in gels or the solutions of polymers. The requirement to develop ecologically clean productions will give an opportunity of wide application of nanoreactors in chemistry and metallurgy. In this case, for the obtaining of metallic nanoparticles and nanowires in carbon shells, it is expedient to evaluate the possibilities of reduction-oxidation couple reactions with the participation of metal ions and organic compounds. The results obtained are experimentally confirmed.

3.2   EXPERIMENTAL AND DISCUSSION

For the synthesis of nanoparticles and nanowires from the mixture of metal salts and polyvinyl alcohol (PVA), the aqueous solutions of salts were mixed in a certain ratio with the aqueous solution of PVA. The average molar ratio of PVA in the mixture was 5. The experiments were carried out on the glass substrates; after the obtained mixtures had been dried, they formed colored transparent films. On some samples, the films were broken due to the large surface tension. The films were heated at 250°C until their color, composition and morphology changed. To control the process, a complex of methods was used, that is, visual spectrophotometry, optical microscopy, X-ray photoelectron spectroscopy and atomic power microscopy.

When PVA was added to the powders of metal chlorides, the color of the mixture changed: the mixture of copper chloride became yellow-green, the cobalt chloride mixture – blue, and nickel chloride – pale-green (Figs. 3.1 and 3.2). Observing the color changes we can conclude that when PVA interacts with metal chlorides the complex compounds are formed.

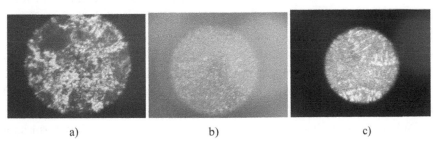

a)   b)   c)

FIGURE 3.1   Photographs of the samples containing PVA and copper chloride (a), cobalt chloride (b), and nickel chloride (c).

Among the aforesaid metals iron is the most active. Brown-red inclusions on the photograph prove the formation of complex iron compounds. In addition, in all the photographs depicting the mixtures containing metal chlorides, we can observe a net of weaves, which are most likely the reflections of nanostructures. The application of optical microscopy and spectrophotometry allows characterizing kinetic parameters of coordination reactions, and also the peculiarities of corresponding reactions because of the possible soiling of raw materials. This is very important for the creation of control conditions during this stage of technological process.

The control of coordination reaction parameters and soiling level gives the possibility evaluates the quality of nanoproduct obtained. As it is shown at other stages of technological process, the quality of nanoproducts and their future application depend upon the characteristics of initial raw materials and intermediate product, which is formed at the first stage.

In order to compare these structures, the morphology of the films changing over a certain range of temperatures was investigated with the help of atomic power microscopy (Fig. 3.4).

When the nanoproduct pictures obtained by atomic power microscopy and optical microscopy are compared with TEM micrograph of the nanoproduct treated thermally and with aqueous solution for the matrix removal, some compliance between them can be noticed. The nanoproduct represents interweaving tubules containing Cu(I), Cu(II). In Fig. 3.3, there are also optical effects indicating light polarization at light transmission through the films owing to the defects appearing during the formation of complex compounds at the initial stage of the process.

Due to the fact that metal ions are active, in the polymer medium they immediately appear in the environment of the PVA molecules and form bonds with hydroxyl groups of this polymer. Polyvinyl alcohol replicates the structure of the particle that it surrounds; however, due to the tendency of the molecules of the metal salts or other metal compounds to combine, PVA seems to envelope the powder particles, and therefore the forms of the obtained nanostructures can be different. The optical microscopy method allows determining the structure of nanostructures at the early stage.

When the samples are heated, metal-containing nanotubes form as a result of dehydration. These processes are thoroughly described [5–6]. Dehydration leads to the darkening of the film. After the samples have been heated, the remaining net of weaves can be seen on the photograph, that is, the structure morphology has remained. To some extent, this fact indicates that the initially formed structure of matrices is inherited. The methods of optical spectroscopy and X-ray photoelectron spectroscopy allow determining the interaction energy of chemical particles in nanoreactors with active centers of nanoreactor walls, which stimulate reduction-oxidation processes (Fig. 3.2).

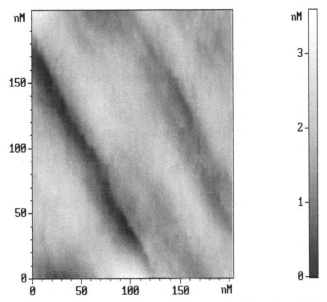

FIGURE 3.2   Micrographs of the surface geometry of PVA film with nickel chloride.

After the heating following the definite program (100–200–300–400°C) the dark brittle products are formed, which are washed out with hot water, dried and milled into dark powders. Nanoproducts obtained are investigated by X-ray photoelectron spectroscopy, transmission and scanning electron microscopy and also electron microdiffraction. For the same samples the Raman spectra are investigated. The main nanoproducts are the tubules (diameter – 20–30 nm), welded with each other (Fig. 3.3). The output of this nanoproduct in the isolated masses after drying exceeds 90%. The yield of carbon-metal-containing nanotubules in account of carbon in PVA equals 85–90%.

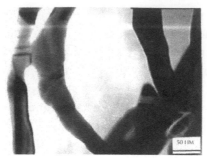

FIGURE 3.3   Micrographs of tubules.

Depending upon the nature of the metal salt and electrochemical potential of the metal, different metal reduction nanoproducts in the carbon shells differing in shape are formed. Based on this result we may speak about a new scientific branch – nanometallurgy.

Based on the results of theoretical and experimental investigations the technological scheme is proposed (*see* Fig. 3.4).

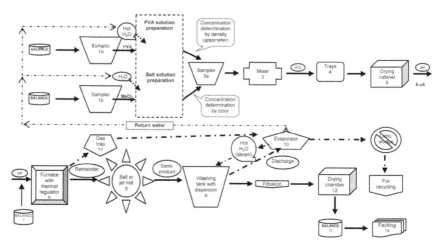

FIGURE 3.4   Technological scheme of nanoproduct production.

The technology proposed was realized with the production of nanoproduct the TEM microphotograph is presented in Fig. 3.5.

FIGURE 3.5   TEM image of nanoproduct obtained in the technology proposed.

Analogous nanoproduct (Ros-4, [6]) is obtained from polymeric materials by high-energy synthesis (Fig. 3.6).

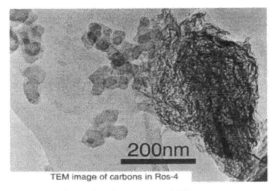

FIGURE 3.6   TEM image of nanoproduct "Ros-4" [6].

However, this nanoproduct (Ros-4) does not contain metals and its price is $25 per gram.

3.3   CONCLUSION

The possible ways for obtaining metallic nanostructures in carbon shells have been determined. The investigation results allow speaking about the possibility of the isolation of metallic and metal-containing nanoparticles in the carbon shells differing in shape and structure. However, still there are problems related to the calculation and experiment because using the existing investigation methods it is difficult to unambiguously estimate the geometry and energy parameters of nanoreactors under the condition of 'erosion' of their walls during the formation of metallic nanostructures in them. The technology of metallic nanoparticles and nanowires in carbon shells is proposed.

KEYWORDS

- nanoproduct
- nanoreactors
- nanowires
- production technology
- Redox processes

REFERENCES

1. V. I. Kodolov, A. P. Kuznetsov, O. A. Nicolaeva et al. Surface and Interface Analysis, 2001, v. 32, pp. 10–14.
2. V. I. Kodolov, I. N. Shabanova, L. G. Makarova et al. J. Struct. Chem., 2001, v. 42, No 2, pp. 260–264 (in Russian).
3. V. I. Kodolov, N. V. Khokhriakov, O. A. Nikolaeva, V. L. Volkov. Chem. Physics and Mesoscopy, 2001, v.3, No 1, pp. 53–65 (in Russian).
4. E. G. Volkova, A.Yu. Volkov, A.M. Murzakaev et al. The Physics of Metals and Metallography, 2003 v.95, No 4, pp. 342–345.
5. A. A. Didik, V. I. Kodolov, A. Yu. Volkov et al. Inorg. Mat., 2003, v. 39, No 6, pp. 693–697.
6. Information of Rosseter Holdings Ltd, in http://www.e-nanoscience.com/products. html; http://www.e-nanoscience.com/process.html; http://www.e-nanoscience.com/ prices.html

PART II
THE PROCESSES MODELING AND COMPUTER SIMULATION

CHAPTER 4

CALCULATION OF THE ELASTIC PARAMETERS OF COMPOSITE MATERIALS BASED ON NANOPARTICLES USING MULTILEVEL MODELING

A. V. VAKHRUSHEV, A. Y. FEDOTOV, and A. A. SHUSHKOV

CONTENTS

ABSTRACT

In this chapter, simulation method of structural, quantitative and deformation properties of nanoparticles and their composites are investigated. The presented model allows to study the dynamic characteristics nano-objects throughout the life cycle of their use, starting from the processes of formation of nanoparticles and ending its influence on the mechanical parameters of the composite.

4.1 INTRODUCTION

The study of the processes of interaction and the formation of substances at the nanoscale allows obtaining new functionality materials, as well as keeping track of their properties and structural features to determine their molecular structure. In this situation the question of formation of nanoparticles, as well as obtaining homogeneous nano-dispersed mixtures of these and other nanoelements for making nanocomposites with homogeneous and stable in terms of the characteristics of the material is given considerable interest (Fig. 4.1). This is due to the fact that the physical-mechanical, chemical and other properties of nanoparticles greatly and typically depend nonlinearly on the nanoparticle size [1–5].

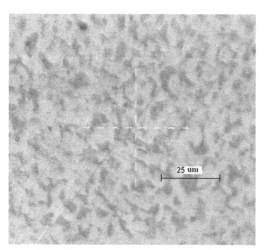

25 um

FIGURE 4.1   The sample surface of the polymer film obtained in plasma benzene steel substrate. Film deposition time is 180 seconds.

The use of nanoparticles is prospectively in environmental terms. The nanoparticles of titanium oxide, and cerium may be dangerous for a person to decompose nitrogen oxides and carbon contained in automobile exhaust. Scientists have conducted research on the treatment of environmental media from contamination by means of nano ingredients. Japanese authors [6] have developed a new nanomaterial, which includes a porous oxide, manganese gold particles grown therein. Made substance effectively removes volatile organic compounds as well as sulfur and nitrogen oxides from air at room temperature.

Composites of nanoparticles are widespread in medicine and pharmacy [7, 8], in which the nanoparticles targeted on drug delivery to cells and protein bodies as well as on artificial muscles and bones. Scientists from the Research Center of American Dental Association in cooperation with the National Institute of Standards and Technology were able to demonstrate that nanotechnology can create sealing material, which is more durable than any previously known therapeutic fillers, and thus it is more effective in preventing repeated failure of a tooth [9]. Authors Quentin le Masne de Chermont, Corinne Chaneac et al. [10] found that some of the nanoparticles containing a metal may be used for visualization of cancer cells. A similar diagnostic effect of heart disease, cancer, and neurodegenerative diseases, but on the basis of nanoparticles containing esters peroxalates and fluorescent dyes, the researchers have made the Georgia Institute of Technology Dongwon Lee, Sirajud Khaja, Juan C. Velasquez-Castano et al. [11]. A team of scientists from the universities of Indiana and Texas have developed the expertise and learned to destroy the tumor cells [12]. Gold nanoshells were injected into the diseased organ, and then delivered to the cancer cells, monocytes and macrophages (immune cells) and under the influence of irradiation led to the elimination of poor quality units (Fig. 4.2).

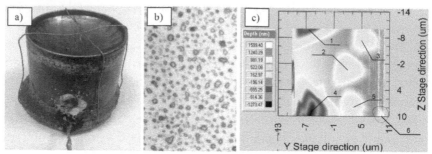

FIGURE 4.2 Extinguishing aerosol generator comprising nanoparticles: a) – appearance generator; b) – view of nanoparticles deposited on the glass after generator operation; c) – section (20 μm × 20 μm) solid precipitated aerosol scanned on a complex system of measurement "Nanotest 600."

Metal nanoparticles have attracted the interest of many researchers because of its functional features [13–16]. Depending on the method of synthesizing nanoparticles of their shape can be varied (Fig. 4.3). Depending on the method of synthesizing nanoparticles of their shape can be varied. There have been cases of synthesis of nanostructures as nanospiraley obtained by the sol-gel method [17], nanofilms deposited by pulsed laser deposition on a substrate [18] and chaotic mnogokonechnyh formations formed by dispersing [19]. Nanoparticles are widespread because the spherical shape of a balanced energy state.

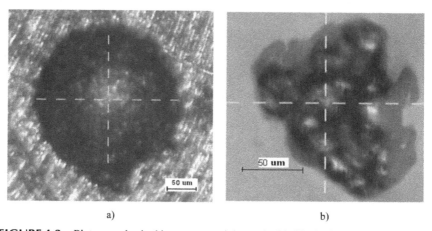

FIGURE 4.3 Pictures spherical iron nanoparticles embedded in the brass plate (a); photos of the iron nanoparticles (b).

Currently, nanotechnology is working closely with many sciences, disciplines and industries. For nanocomposite materials it is necessary to provide uniform and stable in terms of material characteristics, as even a slight variation of the composition, in any local area, could lead to a sharp decrease in the mechanical properties of the nanocomposite that caused the nanodefects under operational loads and significantly reduce product reliability. Experimental study of the mechanisms of formation of nanoparticles is technically difficult and time-consuming task, due to the small size of the data objects. Simulation is an alternative and promising way to study the mechanisms of formation of nano-objects and it is very topical. Determination of physico-mechanical, structural and quantitative characteristics of the size and shape of the nanoparticles formed in order to determine the properties of nanocomposite materials based on its is important.

The aim is to describe a method of modeling the structural, quantitative and deformation properties of nanoparticles and composites based on its throughout the life cycle of nanoparticles starting from forming and finishing phase of use. Value for the practice of the study is that it is related to the calculation of the real process of forming nanoparticles produced according to the modulus of elasticity of the size of the nanoparticles that will allow for the production of micro and nano-composite materials with the desired customer elastic properties. The results can be used as a basis for further investigations of the elastic properties of nano-and microstructural materials of any geometric shape.

4.2   MATHEMATICAL MODEL OF STUDYING NANOPARTICLES

Isolated nanoparticles are generally prepared by evaporation, thermal saturation and the subsequent condensation of the vapor on or near the cold surface [22, 23]. This synthesis technology is easy to use, cost-effective and allows nanocrystalline powders on an industrial scale. Devices using the method of evaporation-condensation, different mode of delivery of the bulk material (powder, solid, liquid), the organization of the heating and cooling of the working environment, the collection of condensed matter. The general principle of operation of such systems is demonstrated in Fig. 4.4.

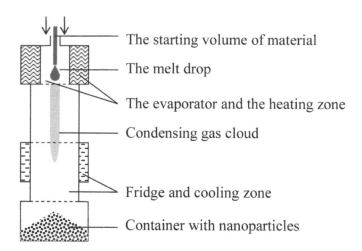

The starting volume of material

The melt drop

The evaporator and the heating zone

Condensing gas cloud

Fridge and cooling zone

Container with nanoparticles

FIGURE 4.4 The scheme of the plant for producing nanoparticles, implementing the principle of thermal saturation and condensation.

Formation nanoelements may occur in vacuum or in an inert gas atmosphere, whereby the particles are rapidly losing kinetic energy due to collisions with gas atoms. High quality nanocomposites using this method is achieved by the high temperature treatment. Cooling the heated composite rate affects the amount of condensation centers, and therefore the formation of nanoparticles and growth rate. An advantage of the evaporation and subsequent condensation is the possibility of using a broad class of nanocomposites.

Simulation of the formation of nanoparticles synthesis technology that simulates the thermal saturation and subsequent condensation was carried out by molecular dynamics [24, 25], which is based on the numerical solution of differential equations of motion Newton for each atom with an initial assignment of velocities and coordinates.

$$m_i \frac{d^2 \bar{\mathbf{r}}_i(t)}{d t^2} = -\frac{\partial U(\bar{\mathbf{r}}(t))}{\partial \bar{\mathbf{r}}_i(t)} + \bar{\mathbf{F}}(\bar{\mathbf{r}}(t), t), \quad i = 1, 2, .., K, \tag{1}$$

$$t = 0, \bar{\mathbf{r}}_i(t_0) = \bar{\mathbf{r}}_{i0}, \frac{d\bar{\mathbf{r}}_i(t_0)}{dt} = \bar{\mathbf{V}}_i(t_0) = \bar{\mathbf{V}}_{i0}, \quad i = 1, 2, .., K, \tag{2}$$

where K – the number of atoms that formed nanosystem; m_i – the mass of the i-th atom; $\overline{r}_{i0}, \overline{r}_i(t)$ – the initial and current radius vector of the i-th atom, respectively; $U(\overline{r}(t))$ – the potential energy of the system; $\overline{V}_{i0},, \overline{V}_i(t)$ – the initial and current speed of the i-th atom, respectively; $\overline{r}(t) = \{\overline{r}_1(t), \overline{r}_2(t), .., \overline{r}_K(t)\}$ – shows the dependence of the location of all the atoms system, $\overline{F}(\overline{r}(t), t)$ – an external force.

Simulation of the formation of nanoparticles was carried out in a representative volume of calculated with periodic boundary conditions. The computational region for the solution of the above problem is shown in Fig. 4.5. A more detailed statement of the problem and the simulation methodology described in earlier papers [26–29].

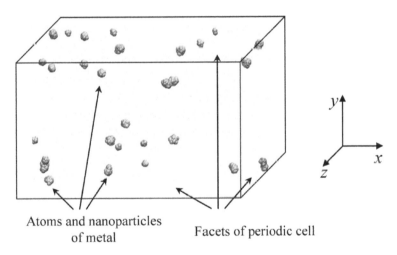

Atoms and nanoparticles of metal   Facets of periodic cell

FIGURE 4.5   The computational region in the simulation of the formation of nanoparticles by thermal saturation and condensation.

4.3 METHOD OF CALCULATION OF NANOPARTICLES ELASTIC PROPERTIES

Method of calculation of the elastic modulus is detailed by the authors [3] and is based on the harmonization of the solutions of the molecular dynamics and the theory of elasticity, carried by vectors of displacements in points coinciding with the position of the atoms of the nanoparticle.

$$\bar{u}_e = \bar{u}_{md} ,$$

(3)

where $\bar{u}_e, \bar{u}_{md}$ – the elastic displacement vectors "equivalent" element and nanoparticles, respectively. Elastic concept of an "equivalent" is based on the same cell response nanoelement elastic member and a predetermined load system. This means that the change in the shape and size of the elastic "equivalent" nanoelement element and must be equal at the same predetermined force system loads.

To satisfy condition (3) is carried out the variation of the elastic constants of the elastic "equivalent" element, so that the total error determined by the difference of displacement vectors were minimal (4).

$$W = \sqrt{\sum_{i=1}^{k} \left(u_{e_i} - u_{md_i} \right)^2} .$$

(4)

Method of calculation of the elastic modulus nanoelements allows us to calculate the elastic parameter specified nanomaterials for arbitrary types of static loading. In the studies involved in the process of forming and producing nanoparticles, it was found that the optimal equilibrium configurations of the particles have a nearly spherical shape (Fig. 4.6). Therefore, as the elastic member is chosen equivalent sphere.

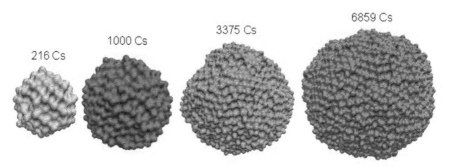

FIGURE 4.6   Equilibrium configuration cesium nanoparticles with different atoms.

The problem of calculating the modulus of elasticity of nanoparticles loaded with axial concentrated forces applied at opposite ends of the diameter is axially symmetric. Therefore, the values of longitudes in this case can be ignored. We denote α as the latitude measured from the radius drawn from the center of the ball to one of the two opposing points of application of concentrated forces on the surface (Fig. 4.7).

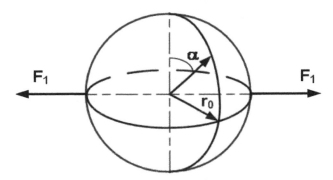

FIGURE 4.7 The elastic "equivalent" element (the ball), stretched concentrated forces.

The problem of calculating the elastic "equivalent" element (the ball) tensile concentrated forces has an analytical solution (5), where r_0 – the radius of the sphere; E – modulus of elasticity; v – Poisson's ratio; u – radial displacements:

$$u = \frac{(1+v)F_1}{2\pi r_0 E} \sum_{n=0}^{\infty} (4n+3) \left\{ (\cos\alpha) \frac{P_{2n+1}(\cos\alpha)}{2n+1} + \right.$$

$$\left. + \left[n(2n+1)\rho^{-2} - n(2n-1) - v(4n+1) \right] \cdot \frac{P_{2n+1}(\cos\alpha)}{m_n} \right\} \rho^{2n+1}, \qquad (5)$$

$$m_n = (1+v)(4n+1)n + n + 1. \qquad (6)$$

$P_{2n}(\cos\alpha)$ – Legendre polynomials; $\rho = r/r_0 < 1$; F_1 – the value of the tensile strength. The function m_n is determined by the expression (6). If the number of terms in the series $n = 48$ function (the Eq. (6)) does not change.

4.4 METHODICS OF CALCULATION OF ELASTIC PROPERTIES OF COMPOSITE MATERIALS

The main deformation characteristics of objects include effective bulk modulus of the material k and the effective shear modulus μ. If necessary, transition is possible from this values to Young's modulus E and Poisson's ratio v according to (7). Backward calculations are possible too.

$$E = \frac{9k\mu}{3k+\mu}, \quad v = \frac{3k-2\mu}{2(3k+\mu)}.$$ (7)

When working with composites and studying its properties, bulk material generally been considered the two components: the matrix and inclusions. The matrix is macro material forming the primary composite structure. The inclusions are some additives in the matrix to refine and improve its mechanical or other characteristics. Deformation properties of the matrix are compatible with the bulk material. They are usually taken from reference books. The strength properties of the inclusions is much more difficult to define. Either experimental data or theoretical methods are used when calculating them. In our case as the inclusions are nanoparticles. To determine the deformation properties and dependencies of the model of the elastic "equivalent" element tensile concentrated forces described above is used.

It is known from previously conducted studies [30, 31] and the effect of the Hall-Petch, the effective bulk modulus and effective shear modulus of nano-objects is not a linear function of their size. Generalized dependence of these quantities can be written in the form of expressions.

$$k_I = \begin{cases} A\left(\dfrac{d}{d_0}\right)^B, \text{if } d \le d_0 \\ k_{I0}, \text{if } d > d_0 \end{cases} = \begin{cases} A\left(\bar{d}\right)^B, \text{if } \bar{d} \le 1 \\ k_{I0}, \text{if } \bar{d} > 1 \end{cases},$$ (8)

$$\mu_I = \begin{cases} M\left(\dfrac{d}{d_0}\right)^L, \text{if } d \le d_0 \\ \mu_{I0}, \text{if } d > d_0 \end{cases} = \begin{cases} M\left(\bar{d}\right)^L, \text{if } \bar{d} \le 1 \\ \mu_{I0}, \text{if } \bar{d} > 1 \end{cases},$$ (9)

where k_I, μ_I – the effective bulk modulus and effective shear modulus of inclusions, $d, \bar{d}$ – actual and relative sizes of the nanoparticles, k_{I0}, μ_{I0} – the effective bulk modulus and effective shear modulus of volume material, which consists of inclusions, d_0 – the diameter of the nanoobjects in which they cease to depend on the elastic properties of the size, A, B, M, L – some of the coefficients, calculated empirically. Having determined the values of the coefficients A, B, M, L, it is possible to calculate the strength parameters of nanostructures of all sizes.

In some cases, it is convenient to use dimensionless or relative deformation characteristics of the inclusions. Dimensionless made in relation to macro properties of material k_{I0} and μ_{I0}:

$$\bar{k}_I = \frac{k_I}{k_{I0}} = \begin{cases} \dfrac{A}{k_{I0}}(\bar{d})^B, \text{if } \bar{d} \le 1 \\ 1, \text{if } \bar{d} > 1 \end{cases}, \quad \bar{\mu}_I = \frac{\mu_I}{\mu_{I0}} = \begin{cases} \dfrac{M}{\mu_{I0}}(\bar{d})^L, \text{if } \bar{d} \le 1 \\ 1, \text{if } \bar{d} > 1 \end{cases}. \quad (10)$$

With a known mechanical parameters and dependencies for inclusions in the form of nanoparticles, the calculation of similar properties of the composite material performed based on of formulas and techniques from Ref. [32]. Model of the medium with low volume fraction of spherical inclusions is expressed by the following relationships:

$$\bar{k} = \frac{k}{k_M} = 1 + \frac{\left(\dfrac{k_I}{k_M} - 1\right)c}{1 + (k_I - k_M)\Big/\left(k_M + \dfrac{4}{3}\mu_M\right)}, \quad \bar{\mu} = \frac{\mu}{\mu_M} = 1 - \frac{15(1 - v_M)[1 - (\mu_I/\mu_M)]c}{7 - 5v_M + 2(4 - 5v_M)(\mu_I/\mu_M)}, \quad (11)$$

where $k, \bar{k}$ – the dimension and the relative effective bulk modulus of the composite, $\mu, \bar{\mu}$ – the dimension and the relative effective shear modulus of the composite, c – the volume fraction of the nanoparticles included in the composite material, k_M, μ_M – the deformation parameters of the matrix.

4.5   RESULTS OF CALCULATION AND ANALYSIS

Formation of composite metal nanoparticles was investigated for the one-, two- and three-component nanoparticles. In all cases, the condensation step in the metal nanoclusters atoms proceeded actively. In the process of grouping involved all types of initial materials. The structure of the nanoobjects was obtained solid. Cavities were not observed in nanoparticles. Nanostructures are obtained predominantly spherical.

Consider the formation of nanoparticles of an example where in starting metals studied ternary mixtures of silver, gold and zinc. The mass fraction of each metal in nanosystem was selected approximately equal to the following values: Ag – 33.97%, Zn – 37.05%, Au – 28.98%. Phase condensation of metal atoms in nanoparticles, following the heat was simulated for 30 ns. The grouping of the atoms in the nanoclusters is actively carried out in the first moments of time and was accompanied by the formation of a significant amount of nanoparticles. Later on, the condensation already formed nanoobjects is observed, which leads to a gradual decrease in the number of nanoparticles and an increase in their size (Fig. 4.8).

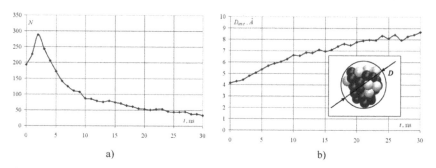

FIGURE 4.8   Changing the number (a) and average diameter (b) of the nanoparticles, formed from 3-component metal mixture in a calculating cell during the condensation phase.

Analysis of the internal structure of layers of nanoparticles was carried out on samples having typical characteristics for all nanoelements. We determined the particle radius and diameter, and then detected the structure and composition of each layer of the nanoparticles as a function of the relative radius of the nanostructure. Graph of the relative density of the layers nanoparticles is shown in Fig. 4.9. The total value of the relative density of each layer was assumed to be 100%. Internal analysis nanoelements showed uneven distribution of metal nanoparticles in the structure under study. The core of the particle consists mainly of gold, the middle layers are formed by atoms of silver, zinc atoms form a shell. There are transition layers in which there are several metals.

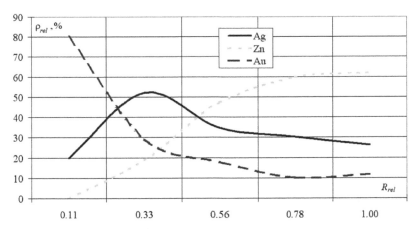

FIGURE 4.9 Change in the relative density of the 3-component nanoparticles depending on the reduced radius.

Detailed use of the proposed methodology for calculating the deformation parameters of a sample of cesium nanoparticles is discussed below. The number of atoms in nanoparticles ranged from 216 to 200,190. Diameter equilibrium nanoparticles with cesium is from 3 nm to 27 nm. The number of atoms is increased as long as the elastic modulus reaches a value nanoparticles reference value Cs. An important task in the development of nanotechnology in Russia and abroad is to determine the size of the nanoparticles, in which the mechanical characteristics will be the same as the reference values of the macro stuff. Of particular interest to the study of mechanical characteristics of the nanoparticles are "small" size nanostructures with the number of atoms of <10,000.

Figure 4.10 (a solid line 1) shows the displacement distance along the loading axis to the center of the "equivalent" elastic element reference values modulus cesium. The same graph (Fig. 4.10, a-2) shows the radial displacements of atoms from the distance to the center of mass of nanoparticles r, for constant Poisson's ratio. The graphs show that these dependencies are not identical. Therefore, changing the modulus of elasticity, it is necessary to achieve data fusion curves (Fig. 4.10b), which is the criterion for the minimum mean square error. The mean square error is a pronounced minimum (Fig. 4.11). We calculate the bulk modulus of nanoparticles of cesium from the system (the Eq. (7)).

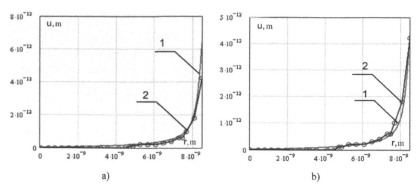

a)   b)

FIGURE 4.10 The dependence of the displacement u from the radius r: 1 – for elastic ball, 2 – for cesium nanoparticles consisting of 49,995 atoms; (a) $E = 1.73 \cdot 10^9$ Pa – reference value of macro material cesium (b) $E = 2.5 \cdot 10^9$ Pa – modulus of elasticity of the investigated nanoparticles.

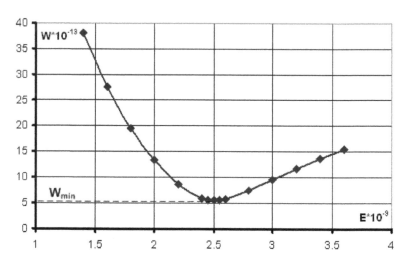

FIGURE 4.11 The dependence of the mean squared error W of the elastic Young's modulus E "equivalent" element, for nanoparticles of cesium (49995 atoms).

Changing the effective bulk modulus dimensionless inclusions depending on the diameter of the above nanoparticles according to Eq. (10) is shown in Fig. 4.12. Triangular markers are shown the data series of numerical experiments, the results of which were subsequently constructed approximating the trend line. The mathematical expression of the curve approximation allowed to determine unknown coefficients $A = 0.882$ and $B = -1.083$ of the Eq. (10). The index of the effective bulk modulus indicates the magnitude of Poisson's ratio corresponding to the reference value of cesium.

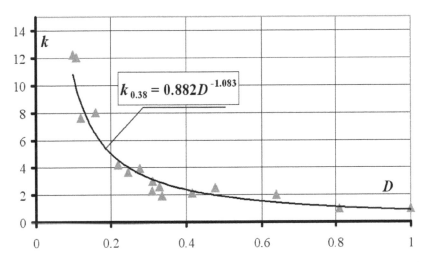

FIGURE 4.12   Changing the dimensionless effective bulk modulus inclusions depending on the dimensionless diameter of inclusions.

The mathematical formulation of the curve of the dimensionless bulk modulus of inclusions is the basis for the calculation of the deformation characteristics of composite materials containing nanoparticles. Bulk modulus and shear modulus of the composite depend on the size of the inclusions contained therein. For a composite material consisting of a polystyrene matrix and nanoinclusions cesium dependence of the dimensionless unit of the reduced diameter is shown in Fig. 4.13.

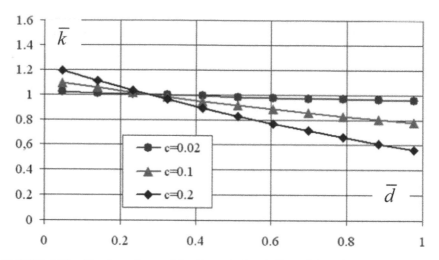

FIGURE 4.13 The dependence of the dimensionless bulk modulus of the nanocomposite of the reduced size of the nanoparticles contained in it.

Analysis of the chart shows that the deformation characteristics are largely dependent on the proportion of the concentration of nanoparticles in the formed composite. Small nanostructures $\left(\bar{d} < 0,3\right)$ increase the bulk modulus of the nanocomposite, large nanoparticles $\left(\bar{d} > 0,3\right)$ lead to its weakening.

4.6 CONCLUSION

Simulation method of structural, quantitative and deformation properties of nanoparticles and composites based on them is offered. The model allows to study the dynamic characteristics nanoobjects throughout the life cycle of their use, starting from the processes of formation of nanoparticles and ending its influence on the mechanical parameters of the composite.

Calculated dependences of the modulus of elasticity, bulk modulus of nanoparticles of cesium on their diameter was obtained. Curves were obtained on the basis of matching the solutions of the molecular dynamics and the theory of elasticity, carried by vectors of displacements in points coinciding with the position of the atoms of the nanoparticle. The critical diameter of nanoparticles cesium (21.5 nm) is calculated. If you increase the size of nanostructures is more critical diameter the strength characteristics of nanoparticles coincide with the reference values of the macro materials.

The dependence of the dimensionless effective bulk modulus of nanoparticles cesium is built. Based on this the mathematical formulation of dependence of the effective bulk modulus is given. Formalization of the change in the effective bulk modulus and shear modulus nanoinclusions depending on their size is an important part to determine the deformation properties of nanocomposite materials.

On the example of polystyrene matrix and cesium nanoparticles it is shown that adding nanostructures in the composite can lead to different effects. Hardening of the material is observed at small sizes of nanostructures. The deterioration of the strength characteristics occurs when the nanoparticles have a significant size.

KEYWORDS

- elastic parameters
- modeling
- nanoparticles
- polystyrene matrix

ACKNOWLEDGMENTS

This work was supported by the Presidium of the Ural Branch of the Russian Academy of Sciences as part of a research project of young scientists 13-1-NP-196 and the Russian Foundation for Basic Research: projects 13-08-01072-a and 11-03-00571-a.

The calculations were performed at the Joint Supercomputer Center of the Russian Academy of Sciences.

REFERENCES

1. Qing-Qing Ni, Yaqin Fu. Masaharu Iwamoto. Evaluation of Elastic Modulus of Nano Particles in PMMA/Silica Nanocomposites Journal of the Society of Materials Science, Japan. 2004. V. 53, №. 9. P. 956–961.
2. Dingreville R., J. Qu, Cherkaoui M. Surface free energy and its effect on the elastic behavior of nano-sized particles, wires and films Journal of the Mechanics and Physics of Solids. 2004. V. 53, №. 8. P. 1827–1854.
3. Vakhrouchev A. V., Shushkov A. A. Method of calculation of the elastic parameters nanoelements Chemical Physics and mesoscopy. 2005. V. 7, №. 3. P. 277–285.
4. Duan H. L., Wang J., Huang Z. P. Size-dependent effective elastic constants of solids containing nano-inhomogeneities with interface stress Journal of the Mechanics and Physics of Solids. 2005. V.53, №.7. P.1574–1596.
5. Gusev A. I., Rempel A. A. Nanocrystalline materials. M.: Fizmatlit. 2000. 224 p.
6. Sinha A.K., Suzuki K., Takahara M. et al. Mesostructured Manganese Oxide-Gold Nanoparticle Composites for Extensive Air Purification Angewandte Chemie Int. Edition. 2007. Is. 16. V. 46, №. 16. P. 2891–2894.
7. Barykinskii G. M., Tuzikov F. V. Physicochemical properties of the silver image hydrosols and other preparations on its basis on the data spectroscopy of surface enhanced Raman scattering spectroscopy, X-ray diffraction analysis and other methods Silver in medicine, biology and engineering. ed. P. P. Rodionova. 1996. P. 136–157.
8. Korenevsky A. V., Solrokin V. V., Karavaiko G. I. The interaction of silver ions with the cells of Candida utilize Microbiology. 1993. V. 62, №. 6. P. 1085–1092.
9. Xu H. H.K., Weir M. D., Sun L. et al. Effects of calcium phosphate nanoparticles on Ca-PO4 composite J. Dent Res. 2007. V. 86(4) P. 378–383.
10. Chermont Q.M., Corinne C., Johanne S. et al. Nanoprobes with near-infrared persistent luminescence for in vivo imaging PNAS. 2007. V. 104, № 22. P. 9266–9271.
11. Dongwon L, Sirajud K., Juan V. C. et al. In vivo imaging of hydrogen peroxide with chemiluminescent nanoparticles Nature Materials. 2007. V. 6. P. 765–769.
12. Ran C., Katie J., Maxy S. et al. A Cellular Trojan Horse for Delivery of Therapeutic Nanoparticles into Tumors Nano Lett. 2007. V. 7. P 3759–3765.
13. Stepanov A. L., Popok V. N., Hole D. L., Bukharaev A. A. The interaction of high-power laser pulses with glasses containing implanted metal nanoparticles Solid State Physics. 2001. V. 43, Issue 11. P. 2100–2106.
14. Koopman M., Gonadec G., Carlisle K., et. al. Compression testing of hollow microspheres (microballoons) to obtain mechanical properties Scripta Materialia. – 2005. – 50. – P. 593–596.
15. Gafner, Yu.A., Gafner S. L. Ni nanoparticles from a gas medium: the appearance and structure Physics of Metals and Metallography. 2005. V. 100, Issue 1. P. 71–76.
16. Vakhrouchev A. V., Severyukhina O. Y., Severyukhin A. V., Vakhrushev A. A., Galkin N. G. Simulation of the processes of formation of quantum dots on the basis of silicides of transition metals Nanomechanics Science and Technology: An International Journal. 2012. Vol. 3, issue 1. P. 51–75.
17. Thomas Delclos, Carole Aim, Emilie Pouget, Aure lie Brizard, Ivan Huc, Marie-Helene Delville and Reiko Oda. Individualized Silica Nanohelices and Nanotubes: Tun-

ing Inorganic Nanostructures Using Lipidic Self-Assemblies Nano Lett., 2008, 8 (7), pp. 1929–1935, DOI: 10.1021/nl080664n.

18. Anais E. Espinal, Lichun Zhang, Chun-Hu Chen, Aimee Morey, Yuefeng Nie, Laura Espinal, Barrett O. Wells, Raymond Joesten, Mark Aindow, Steven L. Suib. Nanostructured arrays of semiconducting octahedral molecular sieves by pulsed-laser deposition Nature Materials, 2010, Volume 9 No 1, pp. 54–59 doi:10.1038/nmat2567.

19. Ruixue Zhang, Louzhen Fan, Yueping Fang and Shihe Yang. Electrochemical route to the preparation of highly dispersed composites of ZnO/carbon nanotubes with significantly enhanced electrochemiluminescence from ZnO J. Mater. Chem., 2008, 18, 4964–4970, DOI: 10.1039/b808769e.

20. Khairul Sozana Nor Kamarudin and Mawarni Fazliana Mohamad. Synthesis of Gold (Au) Nanoparticles for Mercury Adsorption American Journal of Applied Sciences 7 (6): 835–839, 2010.

21. Challa S., R. Kumar Metallic Nanomaterial. WILEY-VCH, 2009, 571 p.

22. Belov V. G., Ivanov V. A., Korobkov, V. A. The apparatus for producing fine metal powders Patent number 2208500. 2003.

23. Vakhrouchev A. V., Fedotov A. Yu., Vakhrouchev A. A. etc. The method and apparatus of mixing nanoparticles Patent number 2301771. 2007. Bull. 18.

24. Riet M. Nanodesign in science and technology. Introduction to the world of nanocalculation. Moscow; Izhevsk: Publishing House of the SIC Regular and Chaotic Dynamics. 2005. 160 p.

25. Heermann D.W. Computer Simulation Methods in Theoretical Physics. Springer-Verlag. 1986. 148 p.

26. Vakhrouchev A. V., Fedotov A.Yu. The study of the laws of probability distribution of the structural characteristics of the nanoparticles simulated by molecular dynamics Computational Continuum Mechanics. 2009. V. 2, Issue 2. P. 14–21.

27. Vakhrouchev A. V., Fedotov A.Yu. The study of the formation of composite nanoparticles from the gas phase by mathematical modeling Chemical Physics and mesoscopy. 2007. Vol. 9, Issue 4. P. 333–347.

28. Vakhrouchev A. V., Fedotov A.Yu. Probabilistic analysis of simulation of the structural characteristics of composite nanoparticles formed in the gas phase Computational Continuum Mechanics. 2008. Vol. 1, Issue 3. P. 34–45.

29. Vakhrouchev A. V., Fedotov A.Yu. Modeling the formation of composite nanoparticles from the gas phase International Scientific Journal for Alternative Energy and Ecology. 2007. Issue 10. Pp. 22–26.

30. Gao H., Ji B., Jager L. I. et al. Materials become insensitive to flaws at nanoscale: Lessons from nature Proc. of the National Academy of Sciences. 2003. V. 100, № 10. P. 5597–5600.

31. Carlton C. E., Ferreira P. J. What is behind the inverse Hall–Petch effect in nanocrystalline materials Acta Materialia. 2007. V. 55. P. 3749–3756.

32. Christensen R. M. Mechanics of composite materials. J. Wiley, 2nd ed. 1979. 348 p.

CHAPTER 5

SIMULATION OF FORMATION OF SUPERFICIAL NANOHETEROSTRUCTURES

A. V. VAKHRUSHEV, A. V. SEVERYUKHIN,
and O. YU. SEVERYUKHINA

CONTENTS

ABSTRACT

In this chapter, a mathematical model describing the processes of formation of heterostructures based on metals and silicon and using the methods of molecular dynamics is suggested. The technique and algorithm of the accounting for the processes of formation and break of chemical bonds in the course of carrying out computational experiments have been created.

5.1   INTRODUCTION

Nanoheterostructures are objects of sizes from one to several hundred nanometers consisting of layers of different semiconductor materials or having inclusions of semiconductor materials in their structure. There are many varieties nanoheterostructures: nanowhiskers, thin nanofilms, quantum dots, super lattices, etc. At the present time, heterostructures are used to create infrared LEDs for the use in fire sensors, perimeter protection systems, night vision devices, infrared illumination in the video surveillance equipment, automatic control systems and other equipment [1, 2]. Controlled synthesis of heterostructures, especially nanoheterostructures of certain composition and morphology, is one of the priorities of modern nanotechnology [3, 4]. The processes of nanoheterostructures are complicated and little studied, therefore computer simulation of the formation of systems based on nanoheterostructural elements is now one of the most important and urgent problems.

This chapter is the development of the researches carried out by authors of Refs. [5–10] on the formation of quantum dots and nanowhiskers on silicon substrate by method of molecular dynamic simulation.

5.2   PROBLEMS AND EQUATIONS

The solution to this problem was carried out by using the developed problem-oriented software [11]. The program is based on the molecular dynamic simulation of the processes that accompany the formation of nanoheterostructures. The use of the methods of molecular dynamics for solving this problem was dictated by the necessity to trace the simulated process kinetics, as well as to evaluate the structure and physical properties of nanoheterostructures. The software includes several units: preparation of initial data, computing unit, data matching unit, the unit of analysis, and visualization of results. Simulation of the formation of nanoheterostructures was carried out using the molecular dynamic method, which describes the motion of a system of atoms, by a system of differential equations. Detailed description of the mathematical models of the formation of quantum dots and nanowhiskers is presented in Refs. [5–10].

5.3 THE ANALYSIS OF THE CALCULATION RESULTS

The process of the formation of nanowhiskers can be divided into several stages [12–14] At the first stage, a thin layer of an active catalyst (e.g., Au) is applied to the surface of the silicon substrate. Then the whole system is heated to a temperature above the eutectic point at which the formation of catalytic agent drops is possible. Further "bombing" of the catalytic agent drops by silicon atoms (Si) is carried out. Silicon atoms are dissolved in the drops. Owing to the Au–Si solution, the supersaturation atoms of silicon under a drop start to crystallize, and nanowhiskers start to form under the gold drops, that is, the growth follows the "vapor–liquid–solid" mechanism.

Particles of the substrate material get to a drop in two ways: directly from the vapor or molecular beam and by diffusion from the lateral surface of the nanowhisker. The diffusive flux from the lateral surface, in turn, consists of:

- particles arriving at the lateral surface of the nanowhisker due to diffusive motion along the surface of the substrate.
- atoms penetrating directly to the sidewalls of the nanowhisker.
- On arriving at the catalytic agent drop, the substrate particles are first dissolved in it, and then crystallize on the surface under the drop; the growth follows the VLS mechanism.

Let us consider explicitly the model of the third stage of the formation of nanowhiskers. As a rule, in the physical models of the growth of nanowhiskers, the chemical potentials of the substances entering into the

considered system is used [15–18]. However, in the method of molecular dynamics it is impossible to directly consider chemical interactions.
It can be done only within quantum-chemical modeling. As mentioned
above, currently it is not obviously possible to calculate large systems by
means of this method. Therefore, to consider the influence of the factors of
chemical interaction, use is made of an additional force operating on each
atom of the system ($\vec{\Phi}_i$), except for the atoms of substrate where $\vec{\Phi}_i = 0$
(Fig. 5.1).

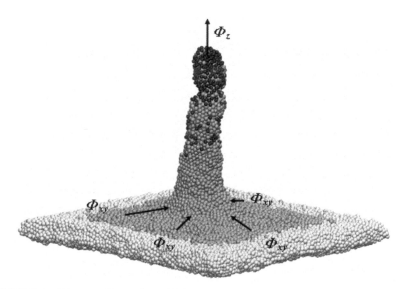

FIGURE 5.1   Diagram illustrating additional forces.

As mentioned above, in order to reflect the actual physical and chemical processes taking place during the growth of nanowhiskers, as well as
to increase their growth rate, an additional force acting on each individual
atom of the system is used [12–14]. In order to explain the nature of this
force, it is necessary to apply main phenomenological models.

During interaction of two phases according to the second law of thermodynamics, their states acquire equilibrium, which is characterized by
the equality of temperatures and pressures of the phases, as well as by
the equality of chemical potentials of each component in the coexisting
phases. The motive force of the transfer of a component from one phase

to another is the difference between the chemical potentials of the component in the interacting phases. The chemical potential characterizes the ability of the component to leave this phase (by evaporation, dissolution, crystallization, chemical reaction, etc.). The chemical potential is incorporated into the first law of thermodynamics and represents the energy of the addition of one particle to the system without performing external work. The transition of the component occurs in the direction of the decrease in its chemical potential. Crystallization is used to separate the crystallizing solid phase from the solution by creating the conditions of supersaturation required for the component.

The achievement of equilibrium by the system is accompanied by the movement of the phase boundary. But this process does not fit within the time frame available for simulation of the molecular dynamics at this stage of development of computer technology, therefore, in simulation, the boundary moves due to the application of force in the direction z (Φ_z). The nature of this force can also be described as the intermolecular interaction.

Diffusion is the process at the molecular level, and it is determined by the random character of the motion of individual molecules. The rate of diffusion is proportional to the average velocity of molecules. It is not obviously possible to track the diffusion process completely in this temporary framework (about several nanoseconds). Therefore, to increase the diffusion rate in accordance with the model of the diffusion–growth of nanowhiskers and Fick's first law (the flux density of matter is proportional to the diffusion coefficient and concentration gradient), an additional diffusional flux to the base of a whisker is used, that is, the force Φ_{xy}, whose direction is perpendicular to the z axis, is applied to the deposited silicon atoms. Correspondingly, this force has the nature of intermolecular interaction.

It is necessary to note that the compound of gold and silicon is formed on the top of the nanowhisker owing to diffusion (Fig. 5.2) it is consistent with the experimental data.

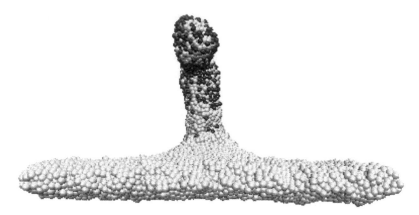

FIGURE 5.2   A mixture of silicon and gold on the top of a whisker (light atoms, Si; dark atoms, Au).

During simulation, the effect of the curling of nanowhiskers was observed in some cases. The emergence of this phenomenon can be explained by the rebuilding of the structure of a nanowhisker, which is in qualitative agreement with experimental data. It is possible to assume that in this case there occurs the formation of the so-called dense packing. Figure 5.3 shows the variation of the internal structure of a nanowhisker in curling. The figure shows that the structure of the nanowhisker being rebuilt. Figure 5.4 also illustrates the rebuilding of the internal structure of the crystal. The selected area is similar to dense packing.

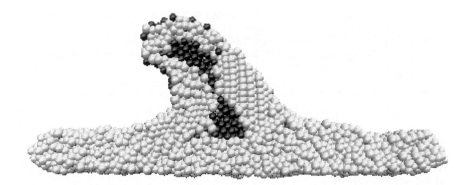

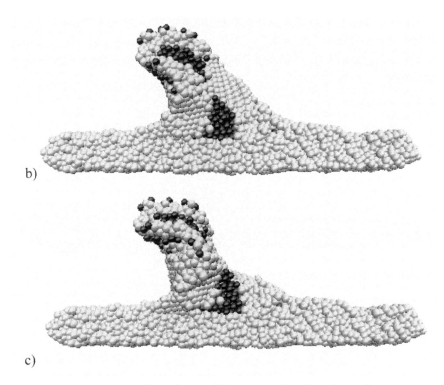

b)

c)

FIGURE 5.3   Stages of curling of nanowhiskers, a section along the x-axis.

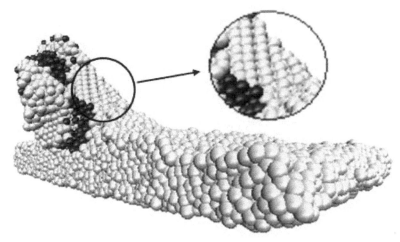

FIGURE 5.4   Rebuilding of the crystal structure.

In some cases after the curling, the melt of Au–Si falls from the peak of the nanowhisker resulting in the so-called silicon nanotips formed on the substrate [19]. Figure 5.5 shows the silicon nanotip obtained in simulation.

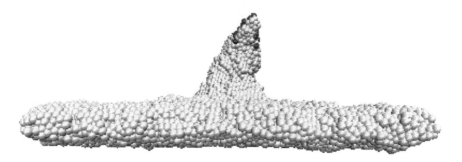

FIGURE 5.5   Nanotip received as a result of simulation.

The plot of the dependence of the height of whiskers on their diameter was presented by authors of Ref. [16] and [20]. During simulation of the process of growth, the results given in Fig. 5.6 were obtained. It is seen from the figure that the characters of the curves representing the dependence of the nanowhisker height on its diameter are similar; in both cases, the dependence of the nanowhisker height on its diameter is decreasing. In Fig. 5.6, the dark dashed line corresponds to the optimal approximation of simulated values; the light line corresponds to the curve L=const/D. In this range of the values of L and D, these functions are almost identical. Table 5.1 shows the average deviation of the approximating function from the theoretical curve L=const/D. It confirms the adequacy and correctness of the proposed mathematical model.

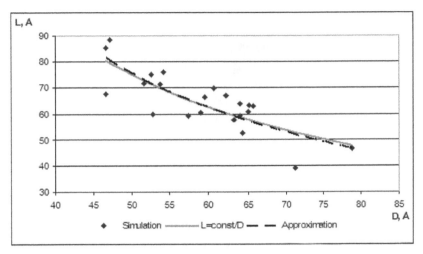

FIGURE 5.6   Dependence of the height of a nanowhisker on the diameter obtained from the results of simulation.

TABLE 5.1   Comparison of the results of simulation with the model proposed by the authors of Ref. [20].

Relative error, %	Absolute error, Å
0.67	0.43

It should be noted that the migration of gold from the drop can occur in silicon nanowhiskers at high temperatures. Similar conclusions can be made from the results of simulation. This phenomenon is also observed in simulation of the processes of the growth of whiskers by the Monte Carlo methods [21] and in our calculations (Fig. 5.7). These data also show a good coordination of simulation results with experiments and other methods of computer simulation.

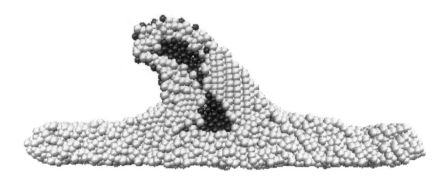

FIGURE 5.7   Gold atoms inside a nanowhisker; a section along the x-axis.

The process of the formation of nanostructures on the substrate surface is greatly influenced by lattice mismatch of deposited material and substrate. There are different mechanisms of growth depending on lattice mismatch. It is known from experimental and theoretical investigations [22] that for the formation of three-dimensional islands it is necessary that the parameter of the mismatch of the lattices in the "deposited material/substrate" system be sufficiently large ($\varepsilon_0 > 2\%$). Moreover, the larger size of the mismatch of lattices the earlier the formation of coherent islands occurs.

For example, the formation of quantum dots is observed in the InAs/GaAs system, where $\varepsilon_0 = 7\%$. In this case, the growth follows the Stranski–Krastanov mode, initially an elastically strained wetting layer, having the same lattice parameter as the substrate material, is formed on the surface. As soon as the wetting layer attains a certain critical thickness, misfit dislocations begin to form. After the formation of such dislocations, the epitaxial layer grows with the lattice constant of the deposited material. For this mechanism, the critical wetting layer thickness exceeds one monolayer.

When $\varepsilon_0 < 2\%$, quantum dots are not formed, and the growth of heterostructures proceeds by the layer-by-layer growth mechanism. At very large values of the mismatch parameter the growth follows the Volmer–Weber mechanism. Then the critical wetting layer thickness is less than one monolayer, and the formation of three-dimensional islands occurs directly on the substrate surface. An example of such a system is the InAs/Si system, where $\varepsilon_0 = 10.6\%$. In a system of silicon and chromium the lattice mismatch parameter is high; therefore here we may also speak of the formation of quantum dots. The increase in the chromium layer thickness up to 0.6–1.5 nm can lead to a change in the morphology of silicide islands. It is noted [23] that since the subsequent deposition of silicon is carried out at a temperature of 1023 K, an increase in the substrate temperature can lead to a change in the size distribution of islands, their coagulation and coalescence with subsequent crystallization due to the increase in the intensity of diffusion processes.

The process of formation of metal–silicon quantum dots is as follows:

1. A silicon substrate is heated to a certain temperature T1.
2. Then, at the same temperature T1 atoms of the metal (Me) are deposited on the substrate surface.
3. Further, system annealing is carried out at temperature T2. At this stage, the formation of islands on the substrate surface occurs. It is known from experimental and theoretical studies that for the formation of three-dimensional islands it is necessary that the lattices mismatch parameter in the deposited material/substrate system be high enough ($\varepsilon_0 > 2\%$). Silicon has a lattice spacing of 5.4307 Å, and the lattice spacing's of most of the metals differ from it significantly. For example, its value for chromium is 2.8850 Å, for iron – 2.866 Å, that is, $\varepsilon_0 > 2\%$ for most of metal–silicon systems. Therefore, in these systems the formation of quantum dots occurs.
4. Silicon atoms are deposited on the obtained system, and as a result the formation of Me–Si quantum dots embedded in silicon occurs.

In studying heterostructures based on silicon and chromium, it was noted that after the deposition of chromium atoms on the surface of Si(111) substrate and with further deposition of silicon, CrSi2 crystallites are formed. Therefore, starting with the second phase (metal deposition) we check the formation of MenSim molecules. If the atoms are located at a distance less than or equal to the equilibrium bond length b_0 (Fig. 5.8), the formation of bonds occurs, the corresponding angles are recalculated, the necessary changes in the configuration files are made. When the atoms that make up a molecule are removed to a distance equal to or greater than r^* (the critical bond length), the bond breaks. Then, the molecular dynamic calculation continues.

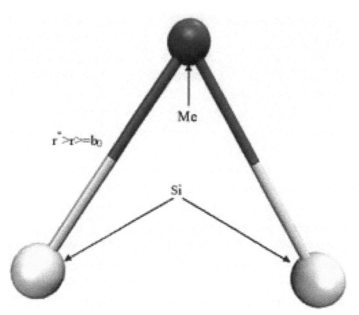

FIGURE 5.8   Formation of bonds (for molecules of the form of MeSi2).

The phenomenon of diffusion exerts a great influence on the morphology and properties of simulated systems. Chromium atoms diffuse into the substrate, but this is observed only in the surface layers, which agrees well with experimental data [23–25]. The presence of diffusion in the surface layers is attributed to the structure and properties of the elements participating in simulation. However, most of the deposited atoms diffuse into the substrate. For example, in the case of deposition of 8000 silicon atoms, 7733 of them diffuse into the substrate, that is, nearly 97%. It can be assumed that precisely these diffusing atoms form CrSi2 molecules. From the experiments it is known that CrSi2 is actively formed during the formation of quantum dots. The process of the adding of bonds is shown in Fig. 5.9.

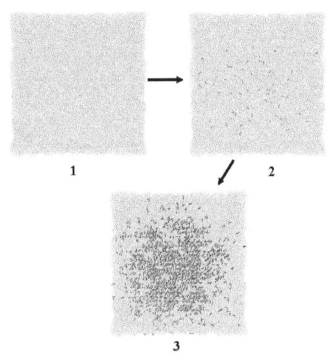

1

2

3

FIGURE 5.9 The process of CrSi2 formation on the substrate surface (top view). The generated CrSi2 molecules are dark in color: 1) $T = 0$; 2) $T = 103$; 3) $T = 105$ fs.

It is known that chromium disilicide belongs to space the group $P6_{2}22$. However, it is rather difficult to verify whether such a structure is formed as a result of simulation. Therefore, the radial pair distribution functions of silicon and chromium atoms were constructed to evaluate the crystallographic structure of the resulting chromium disilicide. Figure 5.10 presents the plots of the radial pair distribution functions for the CrSi2 unit cell and for the system obtained as a result of simulation.

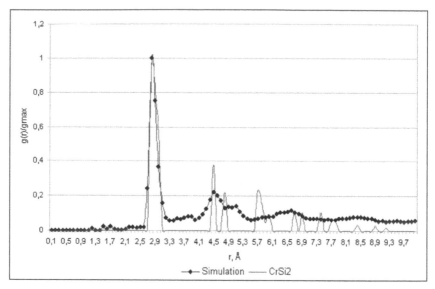

FIGURE 5.10 Radial pair distribution functions (for the CrSi2 unit cell and for the system obtained as a result of simulation).

Proceeding from the good compliance of the profiles of the distribution curves, it is possible to conclude that the simulation results qualitatively correspond to experimental data, and that the $P6_2 22$ structure of CrSi2 is formed in simulation. The quantity of the CrSi2 formed depends on the thickness of the deposited chromium layer and on the formation time.

Annealing of the system and deposition of silicon are carried out at a temperature of 1023 K. As a result of annealing, the formation of bonds occurs more actively, the quantity of CrSi2 increases. This is confirmed by experimental works. In modeling short-term annealing (10 ps in duration), the number of bonds in the simulated system increases on average by 10–15%. For example, for a system with seven monolayers of chromium the quantity of the bonds formed increases from 1036 (at the stage of chromium deposition) up to 1172 (at the stage of annealing), that is, by 13.13%. The active formation of bonds occurs at both the stage of annealing, and the initial stage of silicon deposition (Fig. 5.11).

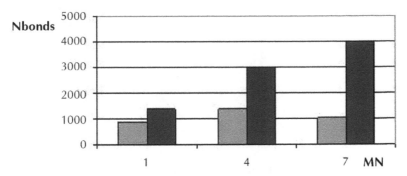

FIGURE 5.11 Effect of the stages of annealing and silicon deposition on the state of the system; N bonds number of bonds; MN, amount of monolayers.

The morphology and properties of heterostructures depend on the thickness of the deposited chromium layer. Moreover, the number of chromium atoms deposited on the substrate surface also affects the structure of the emerging islands. At small thicknesses of chromium layers (1–2 monolayers) almost all the chromium atoms are embedded into the silicon crystal lattice, thus forming CrSi2. However, with increase in the layer thickness up to seven monolayers not all atoms turn out to be built-in and even after the deposition of silicon a certain amount of unbound chromium atoms remain inside the islands.

In some works relating to investigation of the properties and morphology of heterostructures [23–25], it is noted that after the chromium disilicide islands have been overgrown with silicon, the picture corresponding to an atomically pure silicon surface Si (111) 7 × 7 persisted for different thicknesses of chromium. These points to the epitaxial growth of silicon and to the possible embedding of nanoscale heteroepitaxial islands of CrSi2 into the crystal lattice of silicon and their transformation into nanocrystallites mostly spherical in shape. The picture obtained after the deposition of silicon is also consistent with an atomically pure surface, but in some cases, chromium disilicide nanocrystallites, which were not built into the crystal lattice of silicon, emerge onto the surface (Fig. 5.12). The fact that the quantity of CrSi2 emerged onto the surface or located in the surface layer is extremely small indicates that the CrSi2 nanocrystallites remain at a certain depth in the epitaxial layer of silicon, which is confirmed by experimental data [23–25].

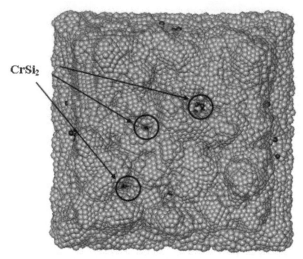

$CrSi_2$

FIGURE 5.12 Simulation results: isolated segments, unembedded CrSi2 nanocrystallites.

5.4 CONCLUSION

A mathematical model describing the processes of formation of hetero-structures based on metals and silicon and using the methods of molecular dynamics is suggested. The technique and algorithm of the accounting for the processes of formation and break of chemical bonds in the course of carrying out computational experiments have been created.

The problem-oriented software package allowing one to realize the solution of is the problem posed has been developed. Complex numerical investigations, which showed the possibility of applying the molecular dynamics method to simulate the formation of quantum dots and the initial stage of the growth of nanowhiskers, are carried out.

The simulation results showed good agreement with experimental data. Thus, the dependence of the diameter of a whisker on its radius is well approximated by the curve L=const/D and has a decreasing character for both experimental data and simulation results. In particular, the formation of three-dimensional islands on the substrate surface is observed at the stages of deposition and annealing. The quantity of CrSi2 emerged onto a surface at the stage of silicon deposition or being in the surface layer is extremely small, pointing to the fact that CrSi2 nanocrystallites remain at a certain depth in the epitaxial layer of silicon.

ACKNOWLEDGMENTS

This work was carried out with financial support from the Research Program of the Ural Branch of the Russian Academy of Sciences: the project 12-P-12–2010.

The calculations were performed at the Joint Supercomputer Center of the Russian Academy of Sciences.

KEYWORDS

- nanoheterostructures
- quantum dots
- simulation
- superficial

REFERENCES

1. Alferov, Zh. I., Andreev, V. M., Garbuzov, D. Z., et al., The influence of heterostructure parameters in the AlAs–GaAs system on the threshold current of lasers and production of continuous lasing at room temperature, Fiz. Tekh. Poluprovodn., vol. 4, pp. 1826–1829, 1970.
2. Alferov, Zh. I., Double heterostructures: Concept and its applications in physics, electronics and technology (Nobel lecture. Stockholm, December 8, 2000), Usp. Fiz. Nauk, vol. 172, no. 9, 2000.
3. Vakhrushev A.V., Fedotov A.Yu., Vakhrushev A.A. Modeling of processes of composite nanoparticle formation by the molecular dynamics technique. Part 1. Structure of composite nanoparticles. Nanomechanics Science and Technology. An International Journal, DOI: 10.1615/NanomechanicsSciTechnolIntJ.v2.i1.20, vol. 2, issue 1, pp. 9–38, 2011.
4. Vakhrushev A.V., Fedotov A.Yu., Vakhrushev A.A. Modeling of processes of composite nanoparticle formation by the molecular dynamics technique. Part 2. Probabilistic laws of nanoparticle characteristics. Nanomechanics Science and Technology. An

International Journal, DOI: 10.1615/NanomechanicsSciTechnolIntJ.v2.i1.30, vol. 2, issue 1, pp. 39–54, 2011.

5. Severyukhina O.Yu., Vakhrushev A.V., Galkin N.G., Severyukhin A.V. Simulation of formation of multilayer nanoheterostructures with variable chemical bonds Chemical physics and mesoscopic. V.13. №.1. 2011. P.53–58.

6. Vakhrouchev A.V., Severyukhina O.Yu., Severyukhin A.V. Simulation of heterogeneous nanostructures formation on silicon surfaces Chemical physics and mesoscopic. V.12. №.2. 2010. P.159–165.

7. Vakhrouchev A.V., Severyukhin A.V., Severyukhina O.Yu. Simulation of heterostructures formation of silicon and chromium Proceedings of the Tula State University. Natural Sciences. I.2. 2011. P.233–240.

8. Vakhrouchev A. V., Severyukhina O. Y., Severyukhin A. V., Vakhrushev A. A., Galkin N. G. Simulation of the processes of formation of quantum dots on the basis of silicides of transition metals. Nanomechanics Science and Technology: An International Journal, vol. 3, issue 1, pp. 51–75, 2012.

9. Vakhrouchev A. V., Severyukhin A. V., Severyukhina O. Y. Modeling the initial stage of formation of nanowhiskers on an activated substrate. Part 1. Theory foundations. Nanomechanics Science and Technology: An International Journal, vol. 3, issue 3, pp. 193–209, 2012. http://scholar.google.com/scholar?btnG=Search%2BScholar&as_q=%22%2BNumerical%2BAnalysis%2Bof%2Bthe%2BAtomic%2Bstructure%2Band%2BShape%2Bof%2BMetal%2BNanoparticles%2B%22&as_sauthors=Vakhrouchev&as_occt=any&as_epq=&as_oq=&as_eq=&as_publication=&as_ylo=&as_yhi=&as_sdtAAP=1&as_sdtp=1

10. Vakhrouchev A. V., Severyukhin A. V., Severyukhina O. Y. Modeling the initial stage of formation of nanowhiskers on an activated substrate. Part 2. Numerical investigation of the structure and properties of au–si nanowhiskers on a silicon substrate. Nanomechanics Science and Technology: An International Journal, vol. 3, issue 3, pp. 211–237, 2012.

11. Vakhrushev, A. V., Severyukhina, O. Yu., and Severyukhin, A. V., The Software Package for the Simulation of the Formation of Quantum Dots, Certificate on Registration of an Electronic Resource no. 17451 as of 09.26.2011, 2011b.

12. Severyukhin, A. V. and Severyukhina, O. Yu., Simulation of the growth of nanowhiskers on an activated substrate, Izv. Tulsk. Gos. Univ., Estestv. Nauki, no. 2, pp. 276–287, 2011.

13. Vakhrushev, A. V., Severyukhin, A. V., and Severyukhina, O. Yu., Modeling of the initial stage of growth of Si–Au nanowhiskers on the Si surface, Khim. Fiz. Mezoskop., vol. 12, no. 1, pp. 24–35, 2010.

14. Vakhrushev, A. V., Severyukhin, A. V., and Severyukhina, O. Yu., Modeling of the growth of heterostructures on the Si substrate, Proc. 2nd All-Russian Conf. "Multiscale Modeling of Processes and Structures in Nanotechnology," Book of Abstracts, Moscow: MIFI Press, p. 377, 2009.

15. Dubrovskii, V. G., Sibirev, N. V., Cirlin, G. E. Shape modification of III-V nanowires: The role of nucleation on sidewalls, Phys. Rev. E, vol. 77, no. 3, pp. 0316061–067, 2008.

16. Dubrovsky, V. G., Theoretical Foundations of the Technology of Semiconductor Nanostructures: Textbook, St. Petersburg: 2006.

17. Dubrovsky, V. G., Sibirev, N. V., Tsyrlin, G. E. Nucleation on the lateral surface and its influence on the shape of filamentary nanowires, Fiz. Tekh. Poluprovodn., vol. 41, no. 10, pp. 1257–1264, 2007.
18. Tsyrlin, G. E., Dubrovsky, V. G., Sibirev, N. V., The diffusion mechanism underlying the growth of GaAs and AlGaAs nanowhiskers in the method of molecular-beam epitaxy, Fiz. Tekh. Poluprovodn., vol. 39, no. 5, pp. 587–594, 2005.
19. Givargizov, E. I., Crystalline nanowhiskers and nanotips, Nature, no. 11, pp. 2003 http: vivovoco.rsl.ru/VV/JOURNAL/NATURE/11_03/NEEDLES.HTM
20. Schubert, L., Werner, P., Zakharov, N. D., [Q7], Silicon nanowhiskers grown on <111> Si substrates by molecular-beam epitaxy, Appl. Phys. Lett., vol. 84, pp. 4968–4971, 2004.
21. Nastovjak, A. G., Neizvestny, I. G., and Shwartz, N. L., Effect of growth conditions and catalyst material on nanowhisker morphology: Monte Carlo simulation, Solid State Phenomena, vols. 156–158, pp. 235–240, 2010.
22. Dubrovskii, V. G., The Theory of Formation of Epitaxial Nanostructures, St. Petersburg: Fizmatlit Press, 2009.
23. Galkin, N. G., Turchin, T. V., and Goroshko, D. L., The influence of the thickness of chromium layer on the morphology and optical properties of heterostructures Si (111)/nanocrystallites CrSi2 (111), Fiz. Tverd. Tela, vol. 50, no. 2, pp. 345–353, 2008.
24. Galkin, N. G, Dozsa, L., Turchin, T. V., et al., Properties of CrSi2 nanocrystallites grown in a silicon matrix, J. Phys.: Condens. Matter., vol. 19, no. 50, pp. 506204–506216, 2007.
25. Galkin, N. G., Goroshko, D. L., Dotsenko, S. A., and Turchin, T. V., Self-organization of CrSi2 nanoislands on Si(111) and growth of monocrystalline silicon with buried multilayers of CrSi2 nanocrystallites, J. Nanosci. Nanotechnol., Amer. Sci. Publ., vol. 8, no. 2, pp. 557–563, 2008.

CHAPTER 6

COMPUTER SIMULATION OF NI, CU AND AU CLUSTER STRUCTURES IN SUPERCOOLED LIQUID STATE

A. V. MURIN

CONTENTS

ABSTRACT

In this chapter, the results of the molecular-dynamics simulation of Ni, Cu and Au in liquid and supercooled liquid states are displayed. The potentials of interatomic interaction within the framework of the embedded-atom method are used to generate realistic atomic configurations. The structural analysis of the cluster structure has been conducted with the use of the bond orientational order parameter of the interatomic bonds. It is shown that the local icosahedral order is present and it enhances at supercooling of the melts of the metals under discussion.

6.1   INTRODUCTION

The investigation of the structure of metal melts has been paid attention to for several decades by now. On the one hand, the interest is due to the complexity of direct investigation of the structure by both experimental and theoretical methods. On the other hand, the understanding of the essence of the processes and the factors influencing them is extremely important for application and is the key for the creation of new materials with unique properties.

The atomic structure of liquid metals evolves at rapid cooling and is inherited in solid state at crystallization or amorphization. Many key properties (electrical, magnetic and mechanical) of material obtained are determined by the combination of long-range translation order and local orientational order. Quasi-crystals [1], which were discovered in 1984, can be used as an example; they are prepared by rapid cooling from melt and show icosahedral symmetry in reciprocal space, which is incompatible with periodicity.

In its turn, at supercooling the evolution of the metal atomic structure will be determined by the character of interatomic interaction and the capability of forming atomic groups with stable bonds (nanosized clusters).

Frank [2] was the first who assumed that the structure of liquid metals could be described on the basis of icosahedral packing. Much later [3] the assumption was confirmed by the molecular-dynamics simulation of a supercooled model melt within the framework of Lennard-Jones potentials.

The main result of the calculations is that they demonstrate an enhancement in icosahedral order at increasing supercooling of melt.

The choice of icosahedron as the first coordination polyhedron in liquid FCC-metals is due to several reasons [4]. First of all, after the FCC-lattice breakage in the process of melting, high coordination number close to 12 retains, and the resulting configuration is in good agreement with 12 vertices of icosahedron. The distortion of cuboctahedron forming the nearest environment of an atom in the FCC-structure into icosahedron is little and atomic displacements make up fractions of the shortest interatomic distance.

In Ref. [5], on the basis of the RDF analysis of liquid Cu, Ag, Au, Fe and Ni and indirect geometrical considerations it is concluded that the BCC-structure elements appear in the structure of the FCC-metal melts. At a larger free volume than that of a FCC-lattice, atomic packing of the BCC-type has minimal dimensions of voids and combines metallic and directional bonds. In addition, in Ref. [5] it is concluded that there are some ordered regions in the melt structure consisting of 1000–2000 atoms of the above metals.

Recently, a surge of publications on the experimental and theoretical investigation of the evolution of local atomic structure in liquid and supercooled liquid states is observed. In Ref. [6], the results of the XAS-experiments on liquid and supercooled copper by the reverse Monte-Carlo method were interpreted. The analysis of the cluster structures obtained indicated the presence of the icosahedral order. In Ref. [7] the liquid Ni atomic structure was simulated by the ab-initio molecular dynamics method. The analysis of the local cluster structure based on the Honeycutt-Andersen method showed the presence of both icosahedral and polyhedral order. In Ref. [8] the ab-initio simulation of liquid and supercooled liquid Cu was conducted. In Ref. [9] the molecular-dynamics simulation of the liquid copper structure and processes of rapid supercooling was carried out as well. For the analysis of the cluster structure, an original method of indices of cluster types was developed and used. The authors report about the presence of icosahedral order and make a conclusion about the presence of some ordered regions with higher density and regions with random atomic packing with lower density in the melt structure.

Thus, the above works indicate that there are signs of icosahedral short-range order even in "simple" single-component metal melts at the temperature of melting and above. However, the above-mentioned results were obtained by different calculation methods and with the use of different sets of physical assumptions and methods for analyzing cluster structure. Consequently, the above works are somewhat disconnected and the results obtained cannot be directly compared.

The objective of this chapter is further comparative study of the evolution of the local cluster structure of single-component d-metals (copper, nickel and gold) in liquid and supercooled liquid states on the basis of one method for conducting calculation experiments and uniform system of assumptions.

This chapter is structured as follows. In Section 1, the calculation details are given. Section 2 is devoted to the analysis and discussion of the results obtained. In Conclusion there is a brief summary of the present investigation.

6.2   CALCULATION METHOD

At present the best method for calculation is the ab-initio molecular-dynamics method allowing simultaneous calculation of the evolution of the atomic system and electron subsystem. In this chapter, however, the classical molecular dynamics method in combination with semiempirical potentials of atomic interaction is used in the framework of the embedded-atom method (EAM) [10]. On the one hand, the EAM-approach proved to be good for the simulation of the metal atomic structure in crystalline and liquid states. On the other hand, the EAM-approach is a reasonable compromise between the calculation complexity and physical validity, which allows to conduct the simulation of a system consisting of a larger number of atoms than that in Refs. [6–9]. In addition, it will allow to establish to what extent the results of the local cluster structure simulation are sensitive to the model describing interatomic bonds.

A cubic crystallite consisting of 864 atoms initially ordered into FCC-lattice was used for the calculations. The simulation was performed at periodic boundary conditions and in the thermodynamic NPT-ensemble. The

EAM-potentials from Ref. [11] were used for gold; potentials from Ref. [12] were used for copper; and potentials from Ref. [13] – for nickel. The chosen EAM-potentials are uniform over the wide range of temperature and density; they provide adequate elastic and energy characteristics of gold, copper and nickel, reasonable thermal extension and temperature of melting of model crystallites.

For the preparation of the model melt, the initial FCC-crystallite was exposed to step-by-step rise of temperature by 150K at $P = 0$. After each change of temperature, the system was relaxed to equilibrium state during 50,000 steps (0.5 fs per step). The transition into liquid state was registered by several criteria. First, the dependence of potential energy on temperature showed sharp jump at the point of melting. Second, the disturbance of crystal order was observed at the visual control over the arrangement of atoms in different crystallographic planes. And, finally, the radial distribution function (RDF) was smeared and became of the character typical for liquid.

Further, the melt models prepared by the above method were exposed to relaxation at certain temperatures which were decreasing step wisely imitating the process of rapid cooling and transition into supercooled state. For copper, the simulation was performed at $T = 1600K$, 1400K, 1350K, 1300K, 1200K, 1100K. At each temperature, the system relaxation was conducted during 100,000 MD-steps (0.5 fs per step). The copper melting point is $T_m = 1365K$; thus, the samples were in liquid state at $T = 1600K$, 1400K, 1350K and in supercooled state at $T = 1300K$, 1200K and 1100K, respectively. For nickel, the simulation was conducted at the temperatures $T = 1900K$, 1800K, 1750K, 1650K, 1600K and 1500K. The nickel melting point is $T_m = 1736K$. For gold, the melting point of which is $T_m = 1336K$, the simulation was conducted at $T = 1700K$, 1536K, 1386K, 1336K, 1286K, 1236K and 1050K. The above temperatures were chosen for the direct comparison with the existing experimental data on the melt structure, the results of the computer simulation of liquid copper and the representative presentation of the melt state in the vicinity of the melting point. At each temperature, the calculation of the time-averaged thermodynamic and structural characteristics of the melt was performed.

6.3   RESULTS AND DISCUSSION

The radial distribution function, $g(r)$, is proportional to the density of atoms at the distance R from a certain atom taken for the central atom. The pair correlator is directly comparable to the structural factor obtained from the experiments on the X-ray scattering and it provides only the information on interatomic distances. Figure 6.1 shows the calculated Cu, Ni and Au RDFs in supercooled state in comparison with the experimental data by Waseda [14]. It should be noted that there is good correspondence for gold. For copper and nickel the correspondence of the calculated and experimental RDFs is satisfactory. The discrepancy can be explained by the inaccuracy of the used interatomic interaction potentials. Nevertheless, the first RDF peak for all metals under study is very well reproduced, which allows to speak about the adequacy of further analysis of the cluster structure of the melts. Moreover, the discrepancy of the calculated and experimental RDFs allows to clarify the degree of the influence of the accuracy of the interatomic interaction description on the cluster structure by the comparison of the results with those in Refs. [7–9], where the exacter ab-initio methods of simulation were used.

Figure 6.1d displays the RDF evolution for the model structure of liquid gold at the supercooling of the melt. An increase in the intensity of the first RDF peak with decreasing temperature indicates the increase of the extent of the short-range order associated with icosahedral and polyhedral local order. However, the given order is not associated with the equilibrium FCC-phase, which will be demonstrated further.

The calculation of average coordination numbers Nc was made by the integration of the first RDF peak. For Ni, the Nc is calculated with the use of the cutoff radius value R_{cut} = 3.40 A which corresponds to the first RDF minimum. For copper, the value R_{cut} = 3.45 A and for gold R_{cut} = 3.80 A are used. The exact determination of the minimum value is complicated, and the change of the value with temperature (*see* Fig. 6.1d) is less than error in its determination. Therefore, a specified value R_{cut} is used for all the temperatures.

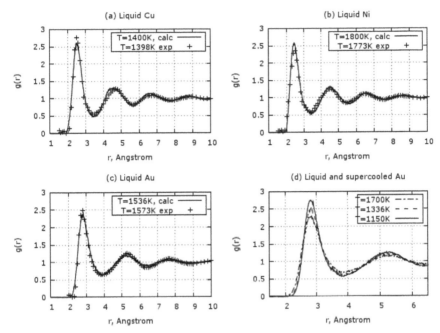

FIGURE 6.1 Comparison of the calculated and experimental functions of radial distribution of supercooled melts: (a) Cu, (b) Ni, (c) Au. The experimental data by Waseda [14] are used. In the insert (d), the evolution of the calculated function of radial distribution for Au at supercooling of the melt is shown.

For nickel, the average value obtained is Nc = 13.0 which practically does not change with temperature. The above results are in good agreement with the results of Ref. [7], where at the increase of the degree of supercooling the Nc values in the range from 12.4 to 13.5 were obtained by the ab-initio molecular dynamics method. The coordination number of the crystalline equilibrium FCC-phase is 12. For copper the values Nc obtained are in the range from 12.5 at the melt temperature 1600K to 13.0 in the supercooled state at 1200K. These results correlate with those in Ref. [8], where the average value Nc = 12.3 was obtained (Nc varies from 12.1 at high temperatures to 12.5 at cooling). For gold the values Nc obtained are in the range from 9.3 for the high temperature melt to 11.0 for the supercooled state. The Nc values obtained are somewhat higher than the experimental ones, however, they show the general tendency, that is, the enhancement of the extent of a certain order at melt supercooling.

At present, there are methods developed which allow to associate a certain quantitative parameter with a certain characteristic atomic surrounding, which unambiguously identifies the given local surrounding and is independent of the variations of angles and distances in a cluster (e.g., ideal icosahedron and distorted icosahedron will have close values for such parameter). Such methods can be exemplified by the methods offered by Honeycutt and Andersen [15] and Steinhardt et al. [3].

Nowadays the problem of unambiguous and unified determination of the presence of interatomic interactions in the structure of rather complex compounds is far from being solved. In our opinion, for the purposes of the current work, the use of the parameters of the orientation order W_l according to the Steinhardt method for the identification of a cluster structure is the most promising, since it will allow the comparative study of different cluster structures on the basis of well-defined quantitative parameters.

For calculating the parameters, the projection of the orientation of bonds from an atom to its nearest neighbors onto the basis of spherical harmonics is performed. The W_l parameter is a rotational invariant of the coefficients in the basis expansion. In a disordered system, the W_l parameters can be calculated for each atom and the distribution function can be formed. The W_l parameter distribution function can be used as a quantitative measure of a local cluster structure.

For ideal icosahedral clusters, $l = 6$ is minimal value for which W_6 is not 0. Table 6.1 enumerates the characteristic values of the W_6 parameter for different ideal atomic clusters. It can be seen that the W_6 value for the icosahedral cluster is far from the values characteristic of other clusters; thus, this parameter is an excellent indicator of the presence of icosahedral order.

For calculating the orientation order parameter, the used cutoff radius R_{cut} values are the same which were used for the calculation of coordination numbers. Figure 6.2 shows the results of the $W6$ parameter calculations for the studied melts at different temperatures.

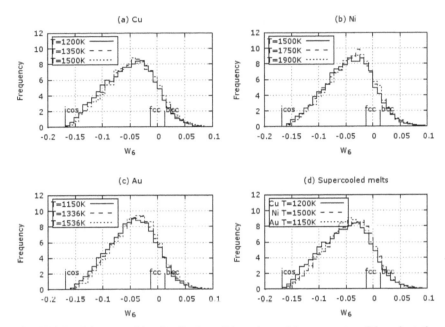

FIGURE 6.2 Diagram of the distribution of the values of the parameter of the orientation order of the interatomic bonds W_6 at different temperature: (a) Cu, (b) Ni, (c) Au. In the insert (d), the comparison of the W_6 distributions is shown for the supercooled melts of the indicated metals.

The values characteristic of the ideal icosahedral, FCC- and BCC-environment are pointed out.

The distribution functions obtained show strong asymmetry towards negative values with tails extending to the values characteristic of ideal icosahedral environment (*see* Table 6.1).

TABLE 6.1 Characteristic values of the W_6 parameter for different atomic clusters.

Cluster	hcp	fcc	icos	Bcc
Number of atoms	12	12	12	14
W_6	−0.012	−0.013	−0.169	+0.013

At $W_6 = -0.169$, the histogram vanishes, which indicates the absence of ideal undistorted icosahedra in the atomic structure. However, it can be noticed that a considerable fraction of clusters have the W_6 value close to the icosahedral value. The above conclusions are valid for all the metals studied in this chapter. In addition, it can be noted that there is a noticeable fraction of clusters in the model of melts, the structure of which is similar to that of the BCC surrounding; it is in good agreement with theoretical models [5].

When the temperature of melt is dropping, the average W_6 value decreases; the maximum of the W_6 parameter distribution shifts to the left. However, the maximum never moves below $W_6 = -0.05$, and the distribution function weakly depends both on temperature and on simulated d-metal. An insignificant increase in the distribution function can be seen on the side of negative values, and, correspondingly, a decrease in the function is observed on the side of positive values, which indicates the insignificant growth of the fraction of clusters with icosahedral and polyhedral order and the decrease of the fraction of clusters with FCC-, HCP- and BCC-like order.

The distribution function obtained is in good agreement both with the results in Refs. [7–9, 16], in which the ab-initio calculation methods were used, and with the results of the simulation by the reverse Monte-Carlo simulation [6]. In addition, it should be noted that the obtained results weakly depend on simulated d-metal (Fe, Ni, Cu, Au). The above allows to conclude that the melt local cluster structure is determined by central short-range repulsive forces to a greater extent and is universal for d-metals with close-packed melting premelting structure.

6.4 CONCLUSION

This study presents the results of the molecular-dynamics simulation of Ni, Cu and Au in liquid and supercooled liquid states using realistic potentials of interatomic interaction. At supercooling of melt, the structure analysis of the evolution of the local cluster structure has been performed on the basis of the parameters of the orientation order of interatomic bonds and the coordination numbers. The presence of local icosahedral and polyhedral order and its insignificant enhancement at the increase of the degree of supercooling are shown. Obviously, such behavior is universal for d-metals with close-packed premelting.

The introduction of the second chemical component into the system, which differs in atomic radius, should lead to a decrease in the degree of the icosahedron distortion by the placement of an atom with a smaller radius in the center and the enhancement of icosahedral order, that is, due to the growth of the number of stable nanosized clusters in the center. The above conclusion is in good agreement with the results of the investigation of the cluster structure of melts [17] showing the presence of nanosized clusters in the melt atomic structure even at high temperatures. In their turn, the icosahedron-like clusters, present in the melt structure, should be the centers of crystallization at rapid cooling and form around themselves a quasi-crystalline phase with local symmetry, which is not compatible with the long-range translation order. Let us note that one of the models describing the structure of quasi-crystals operates with packing of large clusters in 3-d structure [18]. Thus, the results obtained confirm the above theoretical constructs.

The more detailed study of the evolution of the local cluster structure of metals at rapid cooling taking into consideration the presence of the second component is going to be the subject of our further investigation.

ACKNOWLEDGMENT

The work was made possible thanks to the financial support of the Program of Basic Research by Ural Division of the Russian Academy of Sciences (Grant № 12-Y-2-1034).

KEYWORDS

- cluster structure
- embedded-atom method
- liquid metals
- molecular-dynamics simulation

REFERENCES

1. Shechtman, D., I. Blech, D. Gratias, J. W. Cahn. Metallic phase with long-range orientational order and no translational symmetry. Phys. Rev. Lett. 53, n 20, pp. 1951–1953 (1984).
2. F. C. Frank. Supercooling of Liquids. Proc. R. Soc. London, Ser. A 215, N1120, pp. 43–46 (1952).
3. P. J. Steinhardt, D. R. Nelson, M. Ronchetti. Bond-orientational order in liquids and glasses. Phys. Rev. 28, N2, pp. 784–805 (1983).
4. Popel S. I., Spiridonov M. A., Zhukova L. A. Atomic ordering in melted and amorphous metals. Yekaterinburg: UGTU, 1997, 384 p.
5. Pastukhov E. A., Vatolin N. A., Lisin V. L., Denisov V. M., Kachin S. V. Diffraction investigations of the structure of high temperature melts. Yekaterinburg: UrO RAN, 2003. ISBN 5-7691-1308-1, 356 p.
6. A. Di. Cicco, A. Trapananti, S. Fraggioni, A. Filipponi. Is there icosahedral ordering in liquid and supercooled metals?. Phys. Rev. Lett., 91, N13, pp. 135505-1–135505-4 (2003). 7. N. Jakse, A. Pasturel. Ab-initio molecular dynamics simulations of local structure of supercooled Ni. J. Chem. Phys., 120, N13, pp. 6124–6127 (2004).
8. P. Ganesh, M. Widom. Signature of nearly icosahedral structures in liquid and supercooled liquid copper. Phes. Rev. B, 74, N134205, pp. 134205-1–134205-7 (2006).
9. Y. Xue-hua, L. Rang-su, T. Ze-an, H. Zhao-yang, L. Xiao-yang, Z. Qun-yi. Formation and evolution properties of clusters in liquid metal copper during rapid cooling processes. Trans. Nonferrous Met. Soc. China, N18, pp. 33–39 (2008).
10. M. S. Daw, M. I. Baskes. Embedded-atom method: Derivation and application to impurities, surfaces, and other defects in metals. Phys. Rev. B 29, n 12, pp. 6443–6453 (1984).
11. Ercolessi, M. Parinello, E. Tosatti. Simulation of gold in the glue model Philos. Mag. A, 1988, Vol. 58, N1, pp. 213–226.
12. M. I. Mendelev, M. J. Kramer, C. A. Becker, and M. Asta, Analysis of semiempirical interatomic potentials appropriate for simulation of crystalline and liquid Al and Cu. Phil. Mag. 88, pp. 1723–1750 (2008).
13. M. I. Mendelev. http: www.ctcms.nist.gov/potentials/Ni.html.
14. Y. Waseda. The structure of noncrystalline materials. McGraw-Hill. New York (1980).

15. J. D. Honeycutt, H. C. Andersen. Small system size artifacts in the molecular dynamics simulation of homogeneous crystal nucleation in supercooled atomic liquids. J. Phys. Chem. 91, 4950 (1987).
16. P. Ganesh, M. Widom. Ab-initio simulations of geometrical frustration in supercooled liquid Fe and Fe-based metallic glass. Phys. Rev. *B* 77, N1, pp. 014205-1-014205-10 (2008).
17. A. V. Murin, I. N. Shabanova, A. V. Kholzakov. Cluster structure of the A186 Mn14 system in the liquid state. J. Electron Spectroscopy and Related Phenomena, 156–158, pp. 372–374 (2007).
18. M. Mihalkovic, W. -J. Zhu, C. L. Henley, M. Oxborrow. Icosahedral quasi-crystal decoration models: I. Geometrical Principles. Phys. Rev. *B* 53, pp. 9002–9020 (1996).

CHAPTER 7

THE CALCULATING EXPERIMENTS FOR METAL/CARBON NANCOMPOSITES SYNTHESIS IN POLYMERIC MATRIXES WITH THE APPLICATION OF AVRAMI EQUATIONS

V. I. KODOLOV, V. V. TRINEEVA, and YU. M. VASIL'CHENKO

CONTENTS

ABSTRACT

In this chapter, the theoretical ideas about Redox processes in polymeric matrixes nanoreactors are represented. According to these ideas, the Redox process as the main process preceding the nanostructures formation in which the work for charge transport corresponds to the energy of nanocomposite formation process in the reacting layer (nanoreactor). The computational experiment is carried out with the reagents placed in the nanoreactor with the corresponding geometry and energy parameters. Avrami equations are used for such processes which usually reflect the share of a new phase produced. The selection of the modified Avrami equation depends on the nanoreactor shape and size and defines the nanocomposite growth in the nanoreactor. The examples of modified Avrami equation variants for the calculating experiments for the Redox synthesis and the self-organization processes are brought.

7.1   INTRODUCTION

In Refs. [1, 2], the models of nanoreactors in which carbon/metal containing nanostructures can be obtained were considered. The synthesis takes place due to the redox reaction between metal containing substances and nanoreactor walls which are the macromolecules of polymer matrix with functional groups participating in the interaction with metal containing compound or metal ion. If nanoreactors are nanopores or cavities in polymer gels which are formed in the process of solvent removal from gels and their transformation into xerogels or during the formation of crazes in the process of mechanical-chemical processing of polymers and inorganic phase in the presence of active medium, the process essence is as follows:

The nanoreactor is formed in the polymeric matrix which, by geometry and energy parameters, corresponds to the transition state of the reagents participating in the reaction. Then the nanoreactor is filled with reactive mass comprising reagents and solvent. The latter is removed and only the reagents oriented in a particular way stay in the nanoreactor and, if the sufficient energy impulse is available, for instance, energy isolated at the formation of coordination bonds between the fragments of reagents and functional groups in nanoreactor walls, interact with the formation of necessary products. The initially appeared coordination bonds destruct together with the nanoproduct formation.

7.2   THEORETICAL IDEAS ABOUT THE INTERACTIONS IN NANOREACTORS OF POLYMERIC MATRIXES

The difference of potentials between the interacting particles and object walls stimulating these interactions is the driving force of self-organization processes (formation of nanoparticles with definite shapes). The potential jump at the boundary "nanoreactor wall – reacting particles" is defined by the wall surface charge and reacting layer size. If we consider the redox process as the main process preceding the nanostructure formation, the work for charge transport corresponds to the energy of nanoparticle formation process in the reacting layer. Then the equation of energy conservation for nanoreactor during the formation of nanoparticle mol will be as follows:

$$nF\Delta\phi = RT \ln\left(\frac{N_p}{N_r}\right) \tag{1}$$

where n – number reflecting the charge of chemical particles moving inside "the nanoreactor"; F – faraday number; $\Delta\phi$ – difference of potentials between the nanoreactor walls and flow of chemical particles; R – gas constant; T – process temperature; N_p – mol share of nanoparticles obtained; N_r – mol share of initial reagents from which the nanoparticles are obtained.

Using the above equation we can determine the values of equilibrium constants when reaching the certain output of nanoparticles, sizes of nanoparticles and shapes of nanoparticles formed with the appropriate equation modification. The internal cavity sizes or nanoreactor reaction zone and its geometry significantly influence the sizes and shapes of nanostructures.

The sequence of the processes is conditioned by the composition and parameters (energy and geometry) of nanoreactors. To accomplish such processes it is advisable to preliminarily select the polymeric matrix containing the nanoreactors in the form of nanopores or crazes as process appropriate. Such selection can be realized with the help of computer chemistry. Further the computational experiment is carried out with the reagents placed in the nanoreactor with the corresponding geometry and energy parameters. Examples of such computations were given in Ref. [3]. The experimental confirmation of polymer matrix and nanoreactor selection to obtain carbon/metal-containing nanostructures was given in Refs. [4, 5]. Avrami equations are widely used for such processes which usually reflect the share of a new phase produced.

As follows from Avrami equation,

$$1 - \upsilon = \exp\left[-k\tau^n\right], \tag{2}$$

where υ – crystallinity degree, τ – duration, k – value corresponding to specific process rate, n – number of degrees of freedom changing from 1 to 6, the factor under the exponential is connected with the process rate with the duration (time) of the process. Under the conditions of the isothermal growth of the ordered system "embryo," it can be accepted that the nanoreactor activity will be proportional to the process rate in relation to the flowing process.

It was proposed to use the thermodynamics of small systems and Avrami equations to describe the formation processes of carbon nanostructures during recrystallization (graphitization) [6, 7]. These equations are successfully applied [8] to forecast permolecular structures and prognosticate the conditions on the level of parameters resulting in the obtaining of nanostructures of definite size and shape. The equation was also used to forecast the formation of fibers [9]. The application of Avrami equations in the processes of nanostructure formation: a) embryo formation and crystal growth in polymers [8

$$(1 - \upsilon) = \exp\left[-k\tau^n\right], \tag{3}$$

where υ – crystallinity degree, τ – duration, k – value corresponding to the process specific rate, n – number of the degrees of freedom changing from 1 to 6; b) graphitization process with the formation of carbon nanostructures [6, 7].

$$v = 1 - \exp\left[-B\tau^{n}\right], \tag{4}$$

where v – volume share that was changed, τ – duration, B – index connected with the process rate, n – value determining the process directedness; c) process of fiber formation [9].

$$\omega = 1 - \exp\left[-z\tau^{n}\right], \tag{5}$$

where ω – share of the fiber formed, τ – process duration, z – statistic sum connected with the process rate constant, n – number of the degrees of freedom (for the fiber n equals 1).

Instead of k, B or z, based on the previous considerations, the activity of nanoreactors can be used (a).

At the same time, the metal orientation proceeds in interface regions and nanopores of polymeric phase which conditions further direction of the process to the formation of metal/carbon nanocomposite. In other words, the birth and growth of nanosize structures occur during the process in the same way as known from the macromolecule physics [6], in which Avrami equations are successfully used. The application of Avrami equations to the processes of nanostructure formation was previously discussed in the papers dedicated to the formation of ordered shapes of macromolecules [8], formation of carbon nanostructures by electric arc method [6], obtaining of fiber materials [9].

7.3 MODIFIED AVRAMI EQUATION FOR PROCESSES IN NANOREACTORS

Then the share of the product being formed (W) in nanoreactor will be expressed by the following equation,

$$W = 1 - \exp\left(-a\tau^{n}\right) = 1 - \exp\left[-\left(\frac{\varepsilon_{S}}{\varepsilon_{V}}\right)\tau^{n}\right] = 1 - \exp\left\{-\left[\left(\frac{\varepsilon_{S}^{0}d}{\varepsilon_{V}}\right)\frac{S}{V}\right]\tau^{n}\right\}, \tag{6}$$

where a – nanoreactor activity, $a = \varepsilon_{S}/\varepsilon_{V}$; ε_{S} – surface energy reflecting the energy of interaction of reagents with nanoreactor walls, $\varepsilon_{S} = \varepsilon^{0}{}_{S}dS$;

ε_V – nanoreactor volume energy, $\varepsilon_V = \varepsilon°_V V$; $\varepsilon°_s d$ – multiplication of surface layer energy by its thickness, $\varepsilon°_V$ – energy of nanoreactor volume unit, S – surface of nanoreactor walls, V – nanoreactor volume.

When the metal ion moves inside the nanoreactor with redox interaction of ion (mol) with nanoreactor walls, the balance setting in the pair "metal containing – polymeric phase" can apparently be described with the following equation –

$$zF\Delta\phi = RT \ln K = RT \ln\left(\frac{N_p}{N_r}\right) = RT \ln(1-W),\qquad(7)$$

where z – number of electrons participating in the process; $\Delta\phi$ – difference of potentials at the boundary "nanoreactor wall – reactive mixture"; F – Faraday number; R – universal gas constant; T – process temperature; k – process balance constant; N_p – number of moles of the product produced in nanoreactor; N_r – number of moles of reagents or atoms (ions) participating in the process which filled the nanoreactor; W – share of nanoproduct obtained in nanoreactor.

In turn, the share of the transformed components participating in phase interaction can be expressed with the equation which can be considered as a modified Avrami equation,

$$W = 1 - \exp\left[-\tau^n \exp\left(\frac{zF\Delta\phi}{RT}\right)\right],\qquad(8)$$

where τ – duration of the process in nanoreactor; n – number of degrees of freedom changing from 1 to 6. When "n" equals 1, one-dimensional nanostructures are obtained (linear nanostructures, nanofibers). If "n" equals 2 or changes from 1 to 2, flat nanostructures are formed (nanofilms, circles, petals, wide nanobands). If "n" changes from 2 to 3 and more, spatial nanostructures are formed as "n" also indicates the number of degrees of freedom. The selection of the corresponding equation recording form depends on the nanoreactor (nanostructure) shape and sizes and defines the nanostructure growth in the nanoreactor.

In case of nanostructure interaction with the medium molecules, the medium self-organization is effective with the proximity or concordance of the oscillations of separate fragments of nanostructures and chemical bonds of the medium molecules. This hypothesis of nanostructure influence on media self-organization was already discussed [10–13]. At the same time, it is possible to use quantum-chemical computational experiment with the known software products.

According to the Ref. [14] the process vibration nature is discovered. This fact corroborates by IR spectroscopic investigations of nanostructures aqueous soles. Self-organization processes in media and compositions can be compared with the processes of crystalline phase origin and growth. At the same time, the growth can be one-, two- and three-dimensional. For such processes Avrami equations are widely used which usually reflect the share of the new phase appearing. In this case, the degree of nanostructure influence on active media and compositions is defined by the number of nanostructures, their activity in this composition and interaction duration. The temperature growth during the formation of new-phases in self-organizing medium prevents the process development.

Different shapes of nanostructures appear during the formation of lamellar or linear substances. For instance, graphite, clay, mica, asbestos, many silicates have lamellar structure and, consequently, they contain nanostructures with the shape of rotation bodies. It can be explained by the facts that bands of lamellar structure are rolled into clews and spirals with further sewing between them and formation of spheres, cylinders, ellipsoids, cones, etc. Nanostructures formed can be classified based on complexity. The formation of film structures resembling petals, that can be put together into segments and semispheres afterwards, is possible for carbon-containing structures that are formed mainly from hexagons with a certain number of pentagons and insignificant number of heptagons. Under certain energy actions the distortion of coplanar (flat) aromatic rings is known, when π-electrons are shifted and charges are separated on the ring, the ring polarity goes up. For polymeric chains and bands formed from macromolecules the possibilities of formation of super molecular structures, corresponding to the possibility of taking a definite shape, can be predicted with the help of Avrami's equation [8] for one-, two-, and three-dimensional crystallization of macromolecules. Here, energy exchange during the formation of ordered nanostructures and without energy exchange with the surroundings, that is, under athermal and thermal embryo-formation, is considered.

As mentioned before, at two-dimensional growth circles that form petals and further more complex nanostructures resembling flowers can be formed [15]. When being rolled, the circles formed can produce distorted semispheres or can serve as a basis for the transformation into "a beady."

During the redox process connected with the coordination process, the character of chemical bonds changes. Therefore correlations of wave numbers of the changing chemical bonds can be applied as the characteristic of the nanostructure formation process in nanoreactor.

$$W = 1 - \exp\left[-\tau^n\left(\frac{v_s}{v_f}\right)\right], \tag{9}$$

where v_s corresponds to wave numbers of initial state of chemical bonds, and v_f – wave numbers of chemical bonds changing during the process.

The share of nanostructures (W) during the redox process can depend on the potential of nanoreactor walls interaction with the reagents, as well as the number of electrons participating in the process. At the same time, metal ions in the nanoreactor are reduced and its internal walls are partially oxidized (transformation of hydrocarbon fragments into carbon ones). Then the isochoric-isothermal potential (ΔF) is proportional to the product $zF\Delta\varphi$ and Avrami equation will look as follows:

$$W = 1 - k_1 \exp\left[-\tau^n \exp\left(\frac{zF\Delta\phi}{RT}\right)\right], \tag{10}$$

where k_1 – proportionality coefficient taking into account the temperature factor; n – index of the process directedness to the formation of nanostructures with certain shapes; z – number of electrons participating in the process; $\Delta\varphi$ – difference of potentials at the boundary "nanoreactor wall – reactive mixture"; F – faraday number; R – universal gas constant.

Then the equation for defining the share (W) of nanostructures formed can be written down as follows by analogy with the aforesaid equations:

$$1 - W = \exp\left[-\beta a\tau^n\right], \tag{11}$$

where β – coefficient taking into account the changes in the activity in the process of nanoproduct formation.

Substituting the value of a, we get the dependence of the share of nano-structures formed upon the ratio of the energy of nanoreactor internal surface to its volume energy.

$$W = 1 - \exp\left[-\beta\left(\frac{\varepsilon_S}{\varepsilon_V}\right)\tau^n\right] = 1 - \exp\left[-\beta\left(\frac{\varepsilon_S^0}{\varepsilon_V^0}\frac{S}{V}\right)\tau^n\right], \tag{12}$$

If the nanoreactor internal walls become the shells of nanostructures during the process, the nanostructures so obtained are the nanoreactor mirror reflections.

The formation processes of metal-containing nanostructures in carbon or carbon-polymeric shells in nanoreactors can be related to one type of reaction series using the terminology of the theory of linear dependencies of free energies (LFE) [16]. Then it is useful to introduce definite critical values for the volume, surface energy of nanoreactor internal walls, as well as the temperature critical value. When the ration $\lg k/k_c$ is proportional – $\Delta\Delta F/RT$, the ratio W/W_c can be transformed into the following expression:

$$\frac{W}{W_c} = b\exp\left\{-\left(\frac{k}{k_c}\right)\left(\frac{\tau}{\tau_c}\right)^n\right\} = b\exp\left\{-\left(\frac{\tau}{\tau_c}\right)^n \exp\left(-\frac{\Delta\Delta F}{RT}\right)\right\} =$$

$$b\exp\left\{\left(\frac{\tau}{\tau_c}\right)^n\left[\exp k_T k_{VS}\left(\frac{\varepsilon_V}{\varepsilon_{Vk}} - \frac{\varepsilon_S}{\varepsilon_{Sk}}\right)\frac{Q}{T}\right]\right\} \tag{13}$$

Where values with index "c" are correspond to the critical (or standard) values, B – proportionality coefficient considering the temperature factor, k_{VS} – coefficient considering correlations $\varepsilon_V/\varepsilon_{Vk}$ and $\varepsilon_S/\varepsilon_{Sk}$, ε_V and ε_{Vk} – volume energies of nanoreactor and "equilibrium" nanoreactor calculated via the ratios of their volumes; ε_S and ε_{Sk} – surface energy and its equilibrium value, T and θ – temperature of the process and temperature of the equilibrium process; τ – time required to develop the process of nanostructure formation; n – index of the process directedness to the formation of nanostructures of definite shapes. The values of volume and surface energies are given after the transformation of $\Delta\Delta F$ in accordance with Ref. [17], in which the physical sense of Taft constants is substantiated using the indicated energies.

At the same time the share of nanostructures (W) during the redox process can depend upon the potential of nanoreactor wall interaction with reagents, as well as the number of electrons participating in the process. The metal ions in nanoreactor are reduced and its internal walls are partially oxidized (transformation of hydrocarbon fragments into carbon ones).

Then the Helmholtz thermodynamic potential (ΔF) is proportional to the product of $zF\Delta\varphi$ and Avrami equations will be expressed by the following formulae in accordance with the above models, one of them can be as follows:

$$W = 1 - k \exp\left[-\tau^n \exp\left(\frac{zF\Delta\phi}{RT}\right)\right], \tag{14}$$

where k – proportionality coefficient considering the temperature factor, n – index of the process directedness to the formation of nanostructures of definite shapes; z – number of electrons participating in the process; $\Delta\varphi$ – difference of potentials on the border "nanoreactor wall – reaction mixture"; F – Faraday number; R – universal gas constant. When n equals 1, one-dimensional nanostructures are obtained (linear nanosystems and narrow bands). If n equals 2 or changes from 1 to 2, narrow flat nanostructures are formed (nanofilms, circles, petals, broad nanobands). If n changes from 2 to 3 and over, spatial nanostructures are formed, since n also means the number of degrees of freedom. If in this equation we take k as 1 and consider the process in which copper is reduced with simultaneous formation of nanostructures of a definite shape, the share of such formations or transformation degree can be connected with the process duration.

7.4   EXAMPLES OF THE MODIFIED AVRAMI EQUATIONS APPLICATION

The experimental modeling of obtaining nanofilms after the alignment of copper compounds with polyvinyl alcohol at 200°C revealed that optimal duration when the share of nanofilms approaches 100% equals 2.5 h. This corresponds to the calculated value based on the aforesaid Avrami equation. The calculations are made supposing the formation of copper nanocrystals on the nanofilms. It is pointed out that copper ions are predominantly reduced to metal. Therefore it was accepted for the calculations that n equals 2 (two-dimensional growth), potential of redox process during the ion reduction to metal ($\Delta\varphi$) equals 0.34 V, temperature (T) equals 473 K, Faraday number (F) corresponds to 26.81 (A×hour/mol), gas constant R equals 2.31 (W×hour/mol×degree). The analysis of the dimensionality shows the zero dimension of the ratio $\dfrac{zF\Delta\varphi}{RT}$. The calculations are made when changing the process duration with a half-hour increment:

Duration, hours   0.5–1.0 –1.5–2.0–2.5

Content of nanofilms, %   22.5–63.8–89.4–98.3–99.8

If nanofillms are scrolled together with copper nanowires, β is taken as equaled to 3, the temperature increases up to 400°C, the optimal time when the transformation degree reaches 99.97%, corresponds to the duration of 2 h, thus also coinciding with the experiment. According to the calculation results if following the definite conditions of the system exposure, the duration of the exposure has the greatest influence on the value of nanostructure share. The selection of the corresponding equation form depends upon the shape and sizes of nanoreactor (nanostructure) and defines the nanostructure growth in nanoreactor or the influence distribution of the nanostructure on the structurally changing medium. With one-dimensional growth and when the activation zero is nearly zero, the equation for the specific rate of the influence distribution via the oscillations of one bond can be written down as follows:

$$W = 1 - \exp\left[-\beta v \tau^n\right], \tag{15}$$

where v – oscillation frequency of the bond through which the nanostructure influences upon the medium, β – coefficient considering the changes in the bond oscillation frequency in the process. In the case discussed the parameter βv can be represented as the ratio of frequencies of bond oscillations v_{is}/v_{fs}, that are changing during the process. At the same time v_{is} corresponds to the frequency of skeleton oscillations of C–C bond at 1100 cm^{-1}, v_{fs} – symmetrical skeleton oscillations of C=C bond at 1050 cm^{-1}. In this case the equation looks as follows:

$$W = 1 - \exp\left\{-\tau^n \cdot \frac{v_{is}}{v_{fs}}\right\} \tag{16}$$

For the example discussed the content of nanofilms in percentage will be changing together with the changes in the duration as follows:

Duration, hours　　　　　0.5–1.0–1.5–2.0–2.5
Content of nanofilms, %　　23.0–64.9–90.5–98.5–99.9

By the analogy with the above calculations the parameters a in the Eq. (5) should be considered as a value that reflects the transition from the initial to final state of the system and represents the ratios of activities of system states. Under the aforesaid conditions the linear sizes of copper (from ion radius to atom radius) and carbon-carbon bond (from C–C to C=C) are changing during the process. Apparently the structure of copper ion and electron interacts with electrons of the corresponding bonds forming the layer with linear sizes $r_i + l_{C-C}$ in the initial condition and the layer with the size $r_a + l_{C=C}$ in the final condition. Then the equation for the content of nanofilms can be written down as follows:

$$W = 1 - \exp\left\{ -\tau^n \cdot \frac{r_a + l_{C=C}}{r_i + l_{C-C}} \right\} \tag{17}$$

At the same time r_i for Cu^{2+} equals 0.082 nm, r_a for four-coordinated copper atom corresponds to 0.113 nm, bond energy C–C equals 0.154 nm, and C=C bond – 0.142 nm. Representing the ratio of activities as the ratio of corresponding linear sizes and taking the value n as equaled to 2, at the same time changing τ in the same intervals as before, we get the following change in the transformation degree based on the process duration.

Duration, hours　　　　　0.5–1.0–1.5–2.0–2.5
Content of nanofilms, %　　23.7–66.0–91.2–98.7–99.9

Thus, with the help of Avrami equations or their modified analogs we can determine the optimal duration of the process to obtain the required result. It opens up the possibility of defining other parameters of the process and characteristics of nanostructures obtained (by shape and sizes).

The influence of nanostructures on the media and compositions can be assessed with the help of quantum-chemical experiment and Avrami equations.

Modified Avraami equations were tested to prognosticate the duration of the processes of obtaining metal/carbon nanofilms in the system "Cu – PVA" at 200°C [18]. The calculated time (2.5 h) correspond to the experimental duration of obtaining carbon nanofilms on copper clusters.

The influence of nanostructures on the media and compositions was discussed based on quantum-chemical modeling [3]. After comparing the energies of interaction of fullerene derivatives with water clusters, it was found that the increase in the interactions in water medium under the nanostructure influence is achieved only with the participation of hydroxy-fullerene in the interaction. The energy changes reflect the oscillatory process with periodic boosts and attenuations of interactions. The modeling results can identify that the transfer of nanostructure influence onto the molecules in water medium is possible with the proximity or concordance of oscillations of chemical bonds in nanostructure and medium. The process of nanostructure influence onto media has an oscillatory character and is connected with a definite orientation of particles in the medium in the same way as reagents orientate in nanoreactors of polymeric matrixes. To describe this process, it is advisable to introduce such critical parameters as critical content of nanoparticles, critical time and critical temperature [1]. The growth of the number of nanoparticles (n) usually leads to the increase in the number of interaction (N). Also such situation is possible when with the increase of n critical value, n value gets much greater than the number of active nanoparticles. If the temperature exceeds the critical value, this results in the distortion of self-organization processes in the composition being modified and decrease in nanostructure influence onto media.

KEYWORDS

- Avrami equations
- computational experiment
- metal/carbon nanocomposites
- nanoreactor
- polymeric matrixes
- Redox process
- self organization

REFERENCES

1. Kodolov, V. I., Trineeva, V. V., "Perspectives of idea development about nanosystems self- organization in polymeric matrixes," In: Lipanov, A. M., Ed., The problems of nanochemistry for the creation of new materials, Institute for Engineering of Polymer Materials and Dyes, Torun, 2012, pp. 75–100.
2. Shabanova, I. N., Kodolov, V. I., Terebova, N. S., Trineeva, V. V., "X Ray electron spectroscopy in investigation of Metal/Carbon nanosystems and nanostructured materials," Udmurt State University, Moscow-Izhevsk, 2012.
3. Khokhriakov, N. V., Kodolov, V. I., "Quantum-chemical modeling of nanostructure formation," Nanotechnics, No. 2, 2005, pp. 108–112.
4. Kodolov, V. I., A.A. Didik, A.Yu. Volkov, E. Volkova, G., "Low-temperature synthesis of copper nanoparticles in carbon shells," HEIs' news. Chemistry and chemical engineering, Vol. 47, No. 1, 2004, pp. 27–30.
5. Lipanov, A.M. Kodolov, V. I., N.V. Khokhriakov et al., "Challenges in creating nanoreactors for the synthesis of metal nanoparticles in carbon shells," Alternative energy and ecology, No. 2(22), 2005, pp. 58–63.
6. Fedorov, V. B., Khakimova, D. K., Shipkov, N. N., M.A. Avdeenko, "To thermodynamics of carbon materials," Doklady AS USSR, Vol. 219, No. 3, 1974, pp. 596–599.
7. Fedorov, V. B., Khakimova, D. K., Shorshorov M. H. et al., "To kinetics of graphitation," Doklady AS USSR, Vol. 222. No 2, 1975, pp. 399–402.
8. Vunderlikh, B., "Physics of macromolecules, vol. 2," Mir, Moscow, 1979.
9. Serkov, A. T., "Theory of chemical fiber formation," In: Serkov, A. T., Ed., Himiya, Moscow, 1975.
10. Kodolov, V. I., Khokhriakov, N. V., A.P. Kuznetsov et al., "Perspectives of nanostructure and nanosystem application when creating composites with predicted behavior," In: Berlin, A. A., Ed., Space challenges in XXI century, vol. 3, Novel materials and technologies for space rockets and space development, Torus press, Moscow, 2007, pp. 201–205.
11. Krutikov, V. A., A.A. Didik, G.I. Yakovlev et al., "Composite material with nanoreinforcement," Alternative energy and ecology, No. 4(24), 2005, pp. 36–41.
12. Kodolov, V. I., Khokhriakov, N. V., A.P. Kuznetsov. "To the issue of the mechanism of nanostructure influence on structurally changing media during the formation of "intellectual" composites," Nanotechnics, No. 3(7), 2006, pp. 27–35.
13. Khokhriakov, N. V., Kodolov. V. I. "Influence of hydroxyfullerene on the structure of water," Int. J. Quantum Chemistry, Vol. 111, No. 11, 2011, pp. 2620–2624.
14. "Synthesis, functional properties and applications," Proceedings of Conf. NATO-ASI, July August, 2002.
15. Palm, V. A., "Basics of quantitative theory of organic reactions," Himiya, Leningrad, 1967.
16. Kodolov, V. I., "On modeling possibility in organic chemistry," Organic reactivity, Vol. 2, No. 4. 1965, pp. 11–18.
17. Kodolov, V. I., Khokhriakov, N. V., Trineeva, V. V., Blagodatskikh, I. I., "Activity of nanostructures and its display in nanoreactors of polymeric matrixes and in active media," Chem. Phys. & Mesoscopy, Vol. 10, No. 4, 2008, pp. 448–460.
18. Kodolov, V. I., Commentary to the paper (2008) by V.I. Kodolov et al., Chem. Phys. Mesoscopy, Vol. 11, No 1, 2009, pp. 134–136.

CHAPTER 8

ON DIVERSIFIED DEMONSTRATION OF ENTROPY

G. A. KORABLEV, N. G. PETROVA, R. G. KORABLEV,
A. K. OSIPOV, and G. E. ZAIKOV

CONTENTS

ABSTRACT

Various demonstrations of entropy have been discussed, including spatial-energy interactions of atom-molecular systems.

Similarly to the ideas of thermodynamics on the static entropy it is proposed to apply the concept of business quality entropy with the help of which it is possible to evaluate the critical limits of consolidated super-power business structures. The probable formation processes of a business structure are correlated via the entropy of random variables in it.

8.1   INTRODUCTION

The idea of entropy appeared based on the second law of thermodynamics and ideas of the adduced quantity of heat.

The adduced quantity of heat in isothermal process is the relation between the quantity of heat Q obtained by the system and temperature T of heat-release body.

The entropy is the function S of the system state whose differential in the elementary reversible process equals the relation between the infinitely little quantity of heat transferred to the systems and its absolute temperature:

$$ds = \delta Q / T$$

.Using such heat-physical definition we can calculate only the difference between entropies. The entropy itself can only be found with the accuracy to the constant summand (integration constant).

In statistic thermodynamics the entropy of the equilibrious system equals the logarithm of the probability of its definite macrostate:

$$S = k \ln W \tag{1}$$

where W – number of available states of the system or degree of the degradation of microstates; k – Boltzmann's constant.

$$\text{Or: } W = e^{S/K} \tag{2}$$

These correlations are general assertions of macroscopic character, they do not contain any references to the structure elements of the systems considered and they are completely independent from microscopic models [1].

Therefore, the application and consideration of these laws can result in a large number of consequences which are most fruitfully used in statistic thermodynamics.

In statistic thermodynamics the entropy is a function of the system state which helps estimating the process directions and possible changes in them.

At any spontaneous changes in the isolated system the entropy always increases: $\Delta S > 0$

The sense of the second law of thermodynamics comes down to the following:

The nature tends from the less probable states to more probable ones. The most probable is the uniform distribution of molecules through the entire volume. From the macrophysical point, these processes consist in equalizing the density, temperature, pressure and chemical potentials, and the main characteristic of the process is the thermodynamic probability W.

In actual processes in the isolated system the entropy growth is inevitable – disorder and chaos increase in the system, the quality of internal energy goes down.

The thermodynamic probability equals the number of microstates corresponding to the given macrostate.

Since the system degradation degree is not connected with the physical features of the systems, the entropy statistic concept can also have other applications and demonstrations (apart from statistic thermodynamics).

"It is clear that out of the systems completely different by their physical state, the entropy can be the same if their number of possible microstates corresponding to one macroparameter (whatever parameter it is) coincide. Therefore the idea of entropy can be used in various fields. The increasing self-organization of human society ... leads to the increase in entropy and disorder in the environment that is demonstrated, in particular, by a large number of disposal sites all over the earth" [2].

In this research we are trying to apply the concept of entropy to the efficiency of business structures and evaluation of the degree of spatial-energy interactions.

8.2   ENTROPY OF BUSINESS STRUCTURE AGGREGATION

The main properties of capitalistic system providing its economic advantages are: 1) effective competition, and 2) maximal personal interest of each worker.

But on different economy concentration levels these *ab initio* features function and demonstrate themselves differently. Their greatest efficiency corresponds to small business – when the number of company staff is minimal, the personal interest is stronger and competitive struggle for survival is more active. With companies and productions increase the number of staff goes up, the role of each person gradually decreases, the competition slackens as new opportunities for coordinated actions of various business structures appear. The quality of economic relations in business goes down that is entropy increases. Such process is mostly vivid in monostructures at the largest enterprises of large business (syndicates and cartels).

The concept of thermodynamic probability as a number of microstates corresponding to the given macrostate can be modified as applicable to the processes of economic interactions that directly depend on the parameters of business structures.

A separate business structure can be taken as the system macrostate, and as the number of microstates – number of its workers (N) which is the number of the available most probable states of the given business structure. Thus it is supposed that such number of workers of the business structure is the analog of thermodynamic probability as applicable to the processes of economic interactions in business.

Therefore it can be accepted that the total entropy of business quality consists of two entropies characterizing: 1) decrease in the competition efficiency (S_1) and 2) decrease in the personal interest of each worker (S_2), that is, $S = S_1 + S_2$. S_1 is proportional to the number of workers in the company: $S \sim N$, and S_2 has a complex dependence not only on the number of workers in the company but also on the efficiency of its management. It is inversely proportional to the personal interest of each worker. Therefore it can be accepted that $S_2 = 1/\gamma$, where γ – coefficient of personal interest of each worker.

By analogy with Boltzmann's Eq. (1), we have:

$$S = (S_1 + S_2) \sim \left[\ln N + \ln\left(\frac{1}{\gamma}\right) \right] - \ln\left(\frac{N}{\gamma}\right)$$

or $S = k \ln\left(\frac{N}{\gamma}\right),$

where k – proportionality coefficient;

Here n shows how many times the given business structure is larger than the reference small business structure, at which $n = 1$, that is, this value does not have the name.

For nonthermodynamic systems we take $k = 1$. Therefore,

$$S = \ln\left(\frac{N}{\gamma}\right) \tag{3}$$

In Table 8.1, one can see the approximate calculations of business entropy by the Eq. (3) for three main levels of business: small, medium and large. At the same time it is supposed that number n corresponds to some average value from the most probable values. One can see more details in Ref. [3].

When calculating the coefficient of personal interest it is considered that it can change from 1 (one self-employed worker) to zero (0), if such worker is a deprived slave, and for larger companies it is accepted as $\gamma = 0,1–0,01$.

TABLE 8.1 Entropy growth with the business increase.

Structure parameters	Business	
	Small	Average
$N_1 - N_2$	10–50	100–1000
	0.9–0.8	0.6–0.4
S	2.408–4.135	5.116–7.824
	3.271	6.470

Despite of the rather approximate accuracy of such averaged calculations, we can make quite a reliable conclusion on the fact that business entropy, with the aggregation of its structures, sharply increases during the transition from the medium to large business as the quality of business processes decreases. The application of more accurate initial data allows obtaining specific values of business entropy, above which the process of economic relations can reach a critical level.

Actually: "New properties of the system can also accompany the process of its degradation. Under the development we often understand the increase in the system (entropy decrease), but as applicable to the company under order we more often understand regulated hierarchy which cannot be considered as the development attribute. Other development criteria – transition of the system to less probable states and diversity increase including the diversity of possible (potential) states of the systems" [4].

Therefore if crisis phenomena are often followed by the aggregation of business structures – nevertheless this "cannot be considered the development attribute."

And the diversity of business systems is expressed in small and medium business. Therefore the optimal criteria of the more qualitative business are defined by the maximum value of their entropy: $S = 6.47$ (in relative units).

In live systems the entropy growth is compensated via the negative entropy (negoentropy) which is formed through the interaction with the environment. That is a live system is an open one. And business cannot be an isolated system for a long period without the exchange process and interactions with the environment. The role of the external system diminishing the increase in the business entropy must be fulfilled, for example, by the corresponding state and public structures functionally separated from business. The demonopolization of the largest economic structures carried out from the "top" in the evolution way can be the inevitable process here.

But the increase in the personal interest of each worker is defined not only by the parameters of business systems but also depends on the overall arrangement of these processes by the employer. For instance, Ford managed to find such ways of work organization which sharply increased the personal interest of all its employees.

In thermodynamics it is considered that the uncontrollable entropy growth results in the stop of any macro changes in the systems, that is, to their death. Therefore the search of methods of increasing the uncontrollable growth of the entropy in large business is topical. At the same time, the entropy critical figures mainly refer to large business. A simple cut-down of the number of its employees cannot give an actual result of entropy decrease. Thus the decrease in the number of workers by 10% results in diminishing their entropy only by 0.6% and this is inevitably followed by the common negative unemployment phenomena.

Therefore for such supermonostructures controlled neither by the state nor by the society the demonopolization without optimization (i.e., without decreasing the total number of employees) is more actual to diminish the business entropy.

8.3 ENTROPY OF AN ELEMENTARY BUSINESS STRUCTURE

In the process of a new business structure formation we go through seeking and recruiting the personnel, at the same time, the number of personnel of which should correspond to the most probability of this process and for the given system is N_0. Here the key role is played by the probability of random values of this process. The similar picture is also characteristic for informative events: "It appears that for the information characteristics it is also possible to introduce the entropy notion. In the information theory we introduce such a value called the random value entropy:

$$H = \sum_n P_n \log_2 \left(P_n^{-1} \right) \tag{4}$$

Here: value H equals the number of binary digits required for difference (record) of the allowed value of the random value x [2]"; P_n – probability of the appearance of each given record of the random value.

Following the Eq. (4), the entropy of a random value is proportional to the total of probabilities and inversely proportional to the logarithm of their probabilities.

For the characteristic of continuous random value we use the function of probability distribution density [5, 6]:

$$y = f(x) = P(\Delta x_i)/\Delta x_i,$$

where $P(Dx_i)$ – probability of the random value ingress into the interval of its values.

In its sense, the random value entropy is reversely proportional to this function:

$$S_0 \sim 1/y$$

$$\text{or } S_0 \sim \Delta x_i [P(\Delta x)]^{-1} \tag{5}$$

Modifying the Eqs. (4) and (5) and transferring from the binary system to the decimal one, we can have:

$$S_0 = \Delta x_i \ln [P(\Delta x)_i]^{-1} \tag{6}$$

Let us consider the application of the Eq. (6) to the distribution of the probabilities of random processes during the formation of the personnel of an elementary business structure regarding the most probable value N_0.

Each interval of the random values relatively to N_0 can be more or less than its value and equals by the module:

$$\Delta N = |N_0 - N_i|$$

For the probability we take the value $P(x)$:

$$P(\Delta x_i) = \frac{N_0}{N_0 + \Delta N}$$

Then the Eq. (6) looks as follows:

$$S_0 = \Delta N_i \ln \left[\left(\frac{N_0}{N_0 + \Delta N_i} \right) \right]^{-1} \tag{7}$$

The calculations of some points of graphic dependence S_0 on by the Eq. (7) at =20 result in Fig. 8.1.

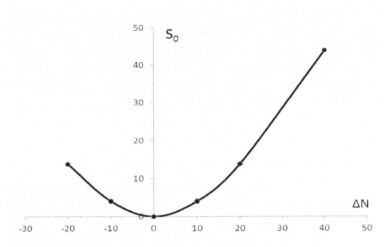

FIGURE 8.1 Entropy of random processes in the formation of the given business structure personnel.

From the calculations and Fig. 8.1, it is seen how the entropy goes up with the deviation of the amount of personnel n from the optimally acceptable value N_0. At the same time, not only those events at which $N < N_0$ are irrational, but also those, at which $N > N_0$. Thus, the nonoptimal increase in the bureaucratic apparatus, which seems to lighten the management work, actually results in such business entropy increase.

Thus, the given technique of entropy evaluation in a separate business structure allows establishing the norms of acceptable deviations from the most probable value of random processes in it.

The use of business entropy idea allows evaluating the quality of business processes, in particular, for the obtaining of the values of their critical parameters.

8.4 ENTROPY OF THE DEGREE OF SPATIAL-ENERGY INTERACTIONS

The idea of spatial-energy parameter (P-parameter) which is the complex characteristic of the most important atomic values responsible for interatomic interactions and having the direct bond with the atom electron density is introduced based on the modified Lagrangian equation for the relative motion of two interacting material points [7].

The value of the relative difference of P-parameters of interacting atoms-components – the structural interaction coefficient is used as the main numerical characteristic of structural interactions in condensed media:

$$\alpha = \frac{P_1 - P_2}{(P_1 + P_2)/2} 100\% \tag{8}$$

Applying the reliable experimental data we obtain the monogram of structural interaction degree dependence coefficient, the same for a wide range of structures (Fig. 8.2). This approach gives the possibility to evaluate the degree and direction of the structural interactions of phase formation, isomorphism and solubility processes in multiple systems, including molecular ones.

Such monogram can be demonstrated [7] as a logarithmic dependence:

$$\alpha = \beta \ln\left(\rho^{-1}\right) \tag{9}$$

where coefficient β – the constant value for the given class of structures. From the average value β can structurally change mainly within $\pm 5\%$. Thus coefficient α is reversely proportional to the logarithm of the degree of structural interactions and therefore can be characterized as the entropy of spatial-energy interactions of atomic-molecular structures.

Actually the more is ρ, the more probable is the formation of stable ordered structures (e.g., the formation of solid solutions), that is, the less is the process entropy. But also the less is coefficient α. Moreover, the values of coefficient α for a high degree of structural interactions (up to 7%) additively coincide with the entropy values for small and medium business. Such regularity in the value changes is also preserved further: for the limited degree of structural interactions coefficient α rises sharply and similarly does the value S during the transition from the medium to large business.

Comparing the monogram (Fig. 8.2) with the data from the Table 8.1, we can see the additivity of business entropy values (S) with the values of the coefficient of spatial-energy interactions (α).

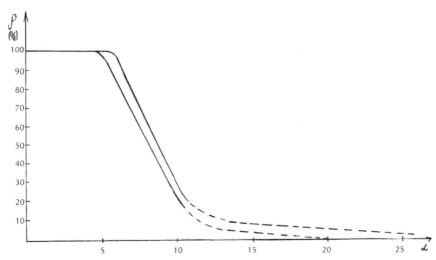

FIGURE 8.2   Dependence of structural interaction degree (ρ) from coefficient α.

Therefore, as applicable to business processes the idea of business quality is similar to the concept of structural interaction degree (ρ). All this allows approximately defining the critical values of these parameters. Thus, at $\rho \approx 10\%$ the value is $S = \alpha \approx 12-18\%$, that corresponds to the number of business structures in the range between 10,000 and 100,000 workers (in the average about 55,000).

The Eq. (9) does not have the complete analogy with Boltzmann's Eq. (1) as in this case not absolute but only relative values of the corresponding characteristics of the interacting structures are compared which can be expressed in percent. This refers not only to coefficient α but also to the comparative evaluation of structural interaction degree (ρ), for example – the percent of atom content of the given element in the solid solution relatively to the total number of atoms.

Therefore in equation (9) coefficient $k = 1$.

8.5   CONCLUSION

The relative difference of spatial-energy parameters of interacting structures can be the numerical characteristic of the entropy of is process.

KEYWORDS

- business entropy
- entropy of a random variable
- spatial-energy parameter
- statistic entropy
- thermodynamic probability

REFERENCES

1. Reif, F. Statistic physics. M.: Nauka, 1972, 352 p.
2. Gribov, L. A., Prokopyeva, N. I. Basics of physics. M.: Vysshaya shkola, 1992, 430 p.
3. Osipov, A. K. Challenges and ways of effective management of a small public catering business. Bulletin of Izhevsk State Agricultural Academy, 2010, №1, p. 29–34.
4. Ivanova, T. Yu., Prikhodko, V. I. Cybernetic-synergetic approach in management theory. Management in Russia and abroad, №5, 2004, p. 132–137.
5. Khrusheva, I. V. Probability theory. Lan Publishers, 2009, 304 p.
6. Liventsev, N. M. Course of physics. Lan Publishers, 2012, 672 p.
7. Korablev, G. A. Spatial-Energy Principles of Complex Structures Formation. Brill Academic Publishers and VSP, Netherlands, 2005, 426 pp. (Monograph).

PART III
SURFACE INVESTIGATION

CHAPTER 9

RESEARCH OF THE LOCAL PHYSICAL AND CHEMICAL STRUCTURE OF SUPERTHIN LAYERS OF POLYMERIC COMPOSITE MATERIALS PROBLEMS AND DECISIONS

S. G. BYSTROV

CONTENTS

ABSTRACT

The new methodology of diagnostics on nano-level of features of a local physical and chemical structure of a surface and interphase layers of the polymeric composite materials is described in the article. The methodology unites opportunities of the methods of XPS, chemical power microscopy and selective chemical reactions with use of some the original and advanced techniques. By means of the developed methodology the data about features of a local physical and chemical structure of a surface and interphase layers of some polymeric composite materials are obtained with the spatial resolution of 20 nanometers. The interrelation of a local physical and chemical structure of the investigated materials with their properties is established and models of formation of the given polymeric compositions are offered.

9.1   INTRODUCTION

In the last few decades, there has been a tendency toward the use of polymeric materials in the areas of science and technology where the manifestation of specific properties of molecular clusters or even individual macromolecules is required: for example, in optoelectronic and data-recording devices, nanoelectronics [1–3] and nanomechanics, and biochemical [4] and biophysical technologies [5, 6].

Development of such devices requires knowledge of the physicochemical surface structure and interphase material layers with depth and area resolutions in the nanometer range. The existing physical methods of research of a surface structure of polymers have a number of restrictions, including the insufficient resolution on the area.

We create the methodology of the diagnostics on a nanosize level of features of a local physical and chemical structure of a surface and interphase layers of polymeric composite materials (PCM), uniting the opportunities of a methods of the X-ray photoelectronic spectroscopy (XPS), chemical power microscopy and selective chemical reactions with use of some the original and advanced techniques. The methodology allows to receive the information on local physical and chemical structure of a surface of polymers (morphology, orientation of macromolecules, mechanical properties) and a local chemical structure of a surface (an element chemical compound, types of the chemical bonds between the atoms on a surface, presence and a spatial arrangement of the certain functional groups).

9.2   RESULTS AND DISCUSSION

Let's result a some examples of the application of the developed methodology:

1.   The morphology of surface PCM studying. The surface of the polymeric films, as a rule, is not exposed to machining treatment, therefore the features of morphology are connected with the processes of a polymers permolecular structures formation.

2.   The mechanical properties of the thin polymeric films measurements. These properties of the given objects are difficult for studying by means of the standard methods. It is offered to use AFM [7]. The some variants are probably:

   a)   The repeated scanning of a film surface site at constant loading. The relative durability of a film is estimated by a quantity of the scans, prior to the beginning of surface destruction.

   b)   The scanning at a gradual increase in the loading at a probe.

   c)   The method of modulation of force (semicontact mode).

   d)   The nanoindentation.

   e)   The measurement of the films thickness by a scratch method.

3.   The processes on the interphase surface investigations by the XPS method. The thin film of a polymer is formed on a filling agent surface for this goal. Then the sample treatments take plaice at the conditions, that was in interest, and the processes on the interphase surface are studied [8].

4.   The cartography of a PCM surface local chemical compound. Following variants are possible:

   a)   The construction of the cards of ACM probe adhesion forces to an investigated surface. Standard probes are used. This method can be applied in that case when components of polymeric system have various force of adhesion to a probe and various hardness [9].

b) The probes modification with the purpose of increase of their selectivity. We offer a method of probes modification in low temperature plasma [10, 11].

c) The selective chemical modification of a PCM surface investigated. This section we shall discuss in more detail.

The method XPS is the most informative technique in studies of the physicochemical surface structure of polymers [12]. It allows obtaining of information on the qualitative and quantitative chemical composition of surfaces virtually without damaging the studied objects. The thickness of the analyzed layer for polymers is no larger than 10 nm. However, for state-of-the-art instruments, the area resolution is 10 μm^2 (when synchrotron radiation is used, a value of 2.5 μm^2 [13] can be reached).

The AFM method ensures surface studies with a high area resolution (in the limit, atomic resolution). To obtain information on the chemical structure of the studied surface by this method, that fact that not only the object topography, but also the nature of force interaction between the tip and surface, contribute to an AFM image is used [14].

In probe microscopy, a trend called chemical force microscopy (CFM) has been formed. To enhance the chemical contrast in AFM images, the methods used in CFM for modifying probes increase their selectivity with respect to the adhesion force to areas of the studied surface with different chemical compositions [15].

However, deposition of self-organizing layers of surface-active substances on the tip is difficult to realize. Moreover, the obtained layers are easily rubbed away and contaminated with fragments of the substance from the surfaces of the studied materials adsorbed on them. This effect affects the reproducibility of results. When adhesive layers are applied, the effective size of the probe tip increases, thus leading to both an error in measurements of the adhesion force, which is due to an increase in the contact area between the tip and the sample surface, and an impaired spatial resolution of obtained AFM images.

We have proposed a technique for studying specific features of the local chemical structure of material surfaces using standard probes by means of the CFM method accompanied by the in situ use of selective chemical reactions (SCRs). Earlier, we have shown that silicon probes possess good selectivity with respect to adhesion to polymers with different surface energies [16]. These differences are especially clearly pronounced in measurements in atmospheres of dried gases. The chemical contrast in AFM images can be intensified via treatment of the studied surface with a reagent that selectively reacts with certain functional groups – the method of SCRs or chemical marks. This method is applied, in particular, in XRPS for recognizing functional groups with equal binding energies in the XPS spectra [17].

The principle of the SCR method is as follows [18, 19]. A sample containing various functional groups, for example, A, B, and C, is treated with reagent–marker xY, which can react only with one functional group present on the surface, for example, C. In this case, a new functional group is produced ("$C-x$"), which contains atoms of the marker element that were previously absent in the sample. After marking, the recognition of the identified functional group in XPS spectra is simplified. Because the content of the marker element in the marked sample is proportional to the number of the recognized functional groups, their number can also be calculated. The force of adhesion of the AFM probe to the surface area that reacted with reagent changes as well.

The technique we developed was used, in particular, for determining the type of functional groups present on the surface of a plasma-polymerized pentane film and their spatial arrangement. As is known from the literature and results of earlier studies, the formation of a film during plasma polymerization begins with the formation of growth centers on the substrate surface [20]. Hence, the surfaces of plasma-polymerized coatings evidently have heterogeneous structures. It is important to determine the spatial position of growth centers and their chemical structure. It was assumed that these centers contain a large number of hydroxyl groups.

A film was deposited in pentane vapors on a substrate made from low-carbon steel in a bell-type reactor using low-temperature high-frequency plasma according to a technique similar to that described in Ref. [21]. A P47 (produced by NT-MDT, Russia) probe microscope was used in AFM investigations. Silicon probes with a 10-nm radius of curvature of tips also produced by NT-MDT were used; the coefficient of elasticity of the probes cantilevers was 0.1 N/m (certified value).

Because we used a method of comparative measurements, it was unnecessary to determine a precise value of the coefficient of elasticity. The forces of adhesion of the probe to the studied surface were determined via measurements of the dependences between the probe– surface interaction force and the probe–surface distance (force spectroscopy) [14].

Measurements were performed in dried nitrogen at 784 points over a mesh of 28 x 28 on the surface of the studied sample with averaging of the results of three measurements. The obtained data were used to plot maps of the adhesion forces. Because the frame size was 1000 nm and the number of pixels in a line during spectroscopy was 28, the spatial resolution during adhesion-force measurements was 1000 nm/28 = 36 nm.

XRPS studies were performed on an ЭC 2401 spectrometer. To process XRPS spectra, we performed calculations according to the additive scheme using fitting by the least-squares method [22, 23]. One plasma-polymerized film and one AFM probe were used in all measurements.

Trifluoroacetic anhydride (TFAA) is the selective reagent for hydroxyl groups [24]. It is a liquid with the low boiling temperature (39.5°C) that contains fluorine atoms. This reagent was used to conduct the SCR.

The schematic diagram of the experiment is shown in fig. 9.1.

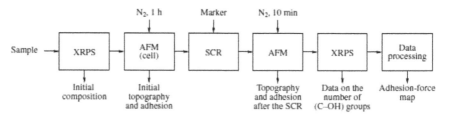

FIGURE 9.1   Schematic of the experiment on the use of SCRs in AFM for obtaining information on the local distribution of hydroxyl groups over the surface of a plasma-polymerized pentane film.

Initially, we obtained the data on the chemical structure of the studied sample surface by means of the XPS method (*see* Table 9.1).

TABLE 9.1 XPS data for the surface of a plasma-polymerized pentane film.

Sample	Analyzed elements					
	carbon (line Cls)		oxygen (line Ols)		fluorine (line Fls)	
	E_b, eV	I_{rel}	E_b, eV	I_{rel}	E_b, eV	I_{rel}
	28.5	1	532.3	1	-	-
Initial	286.9	0.14	533.1	0.17		
	288.2	0.07				
	285	1	532.2	1	688.9	1
	286.9	0.20	534.0	0.41		
After treatment with TFAA vapors	288.1	0.09				
	290.1	0.05				
	292.8	0.07				

Notes: E_b electron binding energy in atoms; I_{rel} relative intessity of lines of the XRPS spectrum.

In the carbon spectrum of the initial film, the e1s line with a binding energy Eb = 285 eV corresponds to aliphatic CH groups of a polymer. The e1s line with Eb = 286.9 eV may be classified as that relating to (C–OH) hydroxyl groups (E_b = 286.7 ± 0.2 eV for the e1s line, and E_b = 533.5 ± 0.2 eV and E_b = 532.7 ± 0.2 eV in the e1s spectrum for aromatic and aliphatic compounds) or simple (C–O–C) ether groups (E_b = 286.5 ± 0.2 eV for the e1s line and E_b = 533.2 ± 0.2 eV and E_b = 532.6 ± 0.2 eV in the e1s spectrum for aromatic and aliphatic compounds). The e1s line with E_b = 288.2 ± 0.2 eV corresponds to (COOH) ester groups of a polymer [25]. The relative intensity of lines is proportional to the concentrations of the corresponding functional groups. Hence, the XRPS data indicate that hydroxyl and ether functional groups are present on the surface of the initial film, thus corresponding to the results from Ref. [20]. However, the XRPS method is unable to determine their local position on the surface of the studied sample.

The sample was placed into a gas–liquid cell mounted on the probe microscope [26]. Dried nitrogen was passed through the cell for 1 h to remove adsorbed water. Then, AFM images of the topography of the studied surface were obtained in a contact mode (Fig. 9.2a).

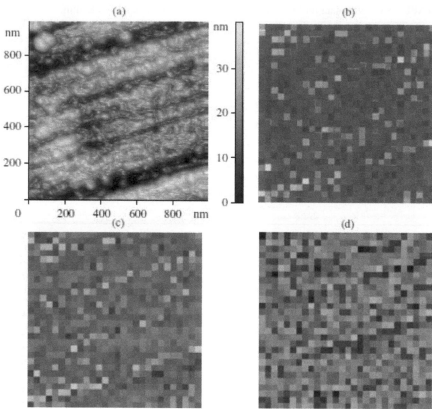

FIGURE 9.2 (a) AFM image of the surface of a plasma-polymerized pentane film in the topography mode; (b) difference map of adhesion forces (the image obtained via subtraction of Fig. 9.2d. from Fig. 9.2c), light regions correspond to surface areas enriched with OH groups; the difference of the adhesion forces change from 0 (dark regions) to 4.2 nN (light regions); (c) map of the forces of adhesion of the probe to the surface of the initial sample, the frame size is 1000 x 1000 nm^2, the adhesion force changes from 3.2 (dark regions) to 7.4 nN (light regions); and (d) map of the forces of adhesion of the probe to the surface of the initial sample after treatment with TFAA, the frame size is 1000 x 1000 nm^2, the adhesion force changes from 3.2 (dark regions) to 7.4 nN (light regions). The pixel size in the maps is 36 x 36 nm. The scanned surface area was 1000 nm^2; the number of points in a line was 512.

Subsequently, the forces of adhesion of the probe to the studied surface were measured by the force spectroscopy method (Fig. 9.3).

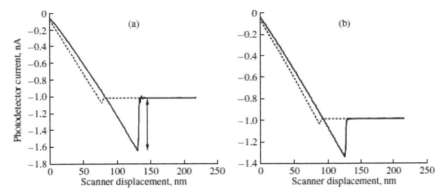

FIGURE 9.3 AFM data for one point (no. 92) of the studied surface of a plasma-polymerized pentane film: (a) before and (b) after treatment of the surface with reagent. Dashed and solid lines correspond to the probe approach and moving away, respectively.

The force of adhesion is proportional to the height of a step in the curve obtained when the probe moves away from the surface (shown by arrows in Fig. 9.3a). The step height is measured in current units (nA or pA) of the microscope photodetector; the obtained values are then converted, using the calibration coefficient and the elasticity coefficient of the probe cantilever, into the values of the adhesion force [14].

After that, the film was treated with TFAA vapors for 1 min. During this procedure, the probe (for protecting the tip against the influence of the reagent) was brought into tight contact with the film surface beyond the area where the adhesion forces were measured. The concentration of TFAA vapors was 0.2 mol/l. Then, reagent vapors were removed from the cell via passage of dried nitrogen for 10 min, and, the sample position being constant, we repeatedly measured the forces of adhesion of the probe to the modified surface (Fig. 9.3b).

The sample was then extracted from the reactor and studied by the XRPS method (*see* Table 9.1). It was found that hydroxyl groups were marked with TFAA, because fluorine atoms with a concentration of 9% appeared in the surface layer of the pentane film. Lines with $E_b = 290.1$ and 292.8 eV, which can be related to carbon atoms in C(O)O and CF_3 groups in TFAA molecules, simultaneously arose in the C1s spectrum [15]. Fluorine appears at the surface owing to the reaction of TFAA with hydroxyl groups:

$$CF_{3\cdot}C(O)\text{-}O\text{-}C(O)\text{-}CF_3 + HO\text{-}C\text{~} \rightarrow CF_{3\cdot}C(O)\text{-}O\text{-}C\text{~} + HO\text{-}C(O)\text{-}CF_3$$

Because, during this reaction, a TFAA fragment containing three fluorine atoms is added to each hydroxyl group, the relative content of carbon atoms included in hydroxyl groups is ~3%. The equation of this reaction shows that six atoms are added to each hydroxyl group. Consequently, the added atoms occupy a larger area on the sample surface than the initial hydroxyls. Thus, a change in the adhesion force after the reaction can be expected on a larger surface area of the studied sample.

On the basis of the force-spectroscopy data, we calculated the average force of adhesion of the probe to the studied surface (data were averaged over the analyzed area) and the adhesion force at each point and maps of the adhesion-force distribution were plotted before and after the sample modification (Figs. 9.2c. and 9.2d, respectively). It has been established that, after marking, the average adhesion force has decreased by 5% as compared to the initial value (5.3 ± 0.03 and 5.6 ± 0.04 nN, respectively). The RMS error of the adhesion-force measurements calculated for each of 784 points resulted in a value no higher than 0.04 nN at a confidence interval of 95%.

Hence, as a result of proceeding SCRs, the force of adhesion of the probe tip to the areas of the studied surface, on which there are hydroxyl groups, decreases. To show these areas more clearly, we graphically subtracted the map shown in Fig. 9.2d. from the map in Fig. 9.2c. In the obtained image (Fig. 9.2b), the areas on which the adhesion force decreased (i.e., hydroxyl groups are concentrated on them) look lighter. As is seen, light areas are positioned mainly along folds on the film surface.

The following facts indicate that the contrast in adhesion-force maps has a physical meaning. The force-spectroscopy data for an individual light point presented (Fig. 9.3). show that, after the chemical reaction was over, the adhesion force at this point decreased by 33%, a value that far exceeds the above measurement error. Here, precisely such points are considered (light dots in the difference map of adhesion forces). As is seen from oscillations of the zero line in (Fig. 9.3)., the instrument noise is no higher than a few picoamperes, whereas the signal (step height) is 600 pA. Therefore, the signal-to-noise ratio is high and the spread of adhesion-force values has a physical nature associated with features of the local structure of the studied surface. It is impossible to improve the contrast for such objects by performing sequential scans, because the polymer surface is destroyed after multiple mechanical influence of the probe tip and the measurement results will be distorted by these processes.

The following arguments can be adduced for discussing the probable contribution of the surface topography to the results of adhesion-force measurements. It is known that the effects of convolution of the images of the probe tip and elements of the surface topography manifest themselves to the maximum extent when their dimensions are close to each other. This rule is also applicable for adhesion-force measurements. If particles of the substance or flaws on the studied surface ave dimensions close to the probe-tip size (10–20 nm for standard tips), in such regions, the area of the contact between the probe and surface increases and the force of adhesion of the probe to the surface correspondingly increases. In our case, surface flaws (folds) have a cross section of ~70 nm. In this case, light dots in the difference map are positioned mainly not in depressions, but along edges of folds; this is most clearly pronounced at the lower edge of images in Fig. 9.2. Thus, although it is impossible to avoid the effect of the topography on the contrast of images observed in the adhesion-force maps, in our case, this effect does not predominate.

In such experiments (with >10-min intervals between repeated measurements), it is important to take into account the drift of the piezoelectric scanner in order to ensure comparison of the obtained images and the adhesion-force maps. We solved this problem, being guided by characteristic details of images of the studied surface. For this purpose, we first obtained an AFM image of the surface with dimensions of 2000 x 2000 nm (Fig. 9.4a) and determined the coordinates of the reference point (shown with an arrow) with an error no larger than ±0.24 nm [27].

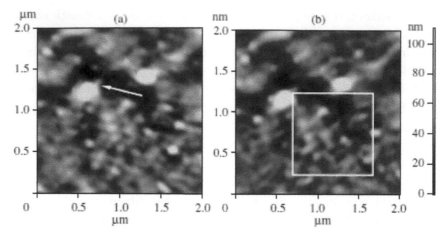

FIGURE 9.4   Use of the reference point for taking into account the image drift: (a) position of the reference point on an area of 2000 x 2000 nm (shown with an arrow); (b) the same area repeatedly recorded 15 min later (the area subjected to spectroscopy is denoted with the square, its upper left corner coincides with the reference point).

Then, in the spectroscopy mode, an area for spectroscopy over a mesh with dimensions of 1000 x 1000 nm was arranged, one of its corners was brought into coincidence with the reference point, and spectroscopic measurements were performed. During the repeated spectroscopy, the area selection procedure was repeated as well (Fig. 9.4b).

9.3   CONCLUSIONS

Hence, the technique we developed allowed us to perform a SCR in the region scanned by a probe microscope and to plot maps of the adhesion forces for the AFM probe to the same polymer surface area before and after its modification by the selective reagent. As a result, we have obtained information on the local chemical structure of the studied plasma-polymerized coating with an area resolution no worse than 36 nm². It has been established that hydroxyl groups on the studied surface are positioned mainly along topographic folds on the film surface [28].

Purposeful controlling of the composition of a gaseous medium during AFM measurements allows one to substantially affect both the quality of

AFM images and the processes proceeding on the substance surface and makes it possible to not only obtain additional information on the studied materials but also create materials with new properties [29].

By means of the developed methodology the data about features of a local physical and chemical structure of a surface and interphase layers of some polymeric composite materials are obtained with the spatial resolution of 20 nanometers. The interrelation of a local physical and chemical structure of the investigated materials with their properties is established and models of formation of the given polymeric compositions are offered.

KEYWORDS

- **a physical and chemical structure**
- **a surface**
- **AFM**
- **interphase layers**
- **polymeric composite materials**
- **XPS**

REFERENCES

1. Fahlman, M. and Salaneck, W.R., Surf. Sci., 2002, vol. 500, nos. 1–3, p. 904.
2. http://www.eurekalert.org/pub_releases/2006–03/rpi-ss 030106.ph
3. http://nanotechweb.org/articles/news/6/6/3/1?rss=2.0
4. http://news.uns.purdue.edu/x2007a/070613bashirSmart nano.htm
5. http://nanotechweb.org/articles/news/5/3/11/1
6. http://nanotechweb.org/articles/news/6/7/5/11
7. Lyakhovitch A.M., Dorfman A.M, Povstugar V.I., Bystrov S.G., Physic of Low-Dimensional Structures, 2001, no 3/4, p.277.
8. Bystrov S.G. et al., Visokomolekularnii coedinenia, 1987, vol. A, no 6, p.1305.
9. Bystrov S.G., Nano-i microsistemnaia technical, 2005, no 6, p.19.
10. Bystrov S.G., Shakov A.A., Zhikharev A.V., Phys. Low-Dim. Struct., 2002, no 5/6, p.47.
11. Patent RU 2381512, no 2008106530.
12. Celotta, R. and Lucatorto, T., Academic Press, 2001, vol. 38, p. 89.

13. Casalis, L., Gregoratti, L., and Kiskinova, M., Surface And Interface Analysis, 1997, vol. 25, p. 374.
14. Butt, H.-J., Cappella, B., and Kappl, M., Surface Science Reports, 2005, vol. 59, p. 1.
15. Duwez, A.-S., Poleunis, C., and Bertrand, P., Langmuir, 2001, vol. 17, p. 6351.
16. Bystrov, S.G., Shakov, A.A., and Zhikharev, A.V., Phys. Low-Dim. Struct., 2002, vol. 5/6, p. 47.
17. Briggs, D. and Seah, M.P., Practical Surface Analysis by Auger and X-ray Photoelectron Spectroscopy, Chichester: Wiley, 1983.
18. Siggia, S. and Hanna, J.G., Quantitative Organic Analysis via Functional Groups, New York: Wiley, 1979.
19. Povstugar, V.I., Mikhailova, S.S., and Shakov, A.A., Zh. Anal. Khim., 2000, vol. 55, no. 5, p. 455 [J. Anal. Chem. (Engl. Transl.), vol. 55, no. 5, p. 405–416].
20. Dorfman, A.M., Lyakhovich, A.M., and Povstugar, V.I., Zashch. Met., 2003, vol. 39, no. 1, p. 70.
21. Bystrov, S.G., Dorfman, A.M., Lyakhovich, A.M., and Povstugar, V.I., Poverkhnost, 2000, no. 11, p. 40.
22. Nefedov, V.N., Rentgenoelektronnaya spektroskopiya khimicheskikh soedinenii (X-ray Photoelectron Spectroscopy of Chemical Compounds), Moscow: Khimiya, 1984.
23. Proctor, A. and Sherwood, P.M.A., Anal. Chem., 1982, vol. 54, p. 13.
24. Chillkoti, A., Ratner, B.D., and Briggs, D., Chem. Mater., 1991, vol. 3, p. 51.
25. Beamson, G. and Briggs, D., High Resolution XPS of Organic Polymers, Chichester: Wiley, 1992, p. 277.
26. Zhikharev, A.V. and Bystrov, S.G., Prib. Tekh. Eksp., 2004, no. 6, p. 116.
27. TU 4254–003–58699387–2004, Dopolnitel'nye obshchie trebovaniya dlya skanerov (Additional general Requirements for Scanners).
28. Bystrov S.G., Instruments and Experimental Techniques, 2009, vol. 52, no. 2, p. 295.
29. Bystrov, S.G., Khimicheskaya Fizika i Mezoskopiya, 2008, vol. 10, no. 1, p. 37.

CHAPTER 10

STRUCTURE AND SURFACE LAYER PROPERTIES OF MEDIUM CARBON STEEL AFTER ELECTROEXPLOSIVE COPPER PLATING AND SUBSEQUENT ELECTRON-BEAM PROCESSING

V. YE. GROMOV, YU. F. IVANOV, D. A. ROMANOV, G. TANG,
S. V. RAYKOV, E. A. BUDOVSKIKH, and G. SONG

CONTENTS

ABSTRACT

The surface relief and structure peculiarities of steel 45 (C≤0.45 wt %) after electroexplosive copper plating and subsequent electron beam treatment are investigated by methods of scanning and transmission electron microscopy. It is established that the copper concentration in surface layer is increased in two times with the growth of electron beam pulses number. The high speed crystallization on of modified layer is accompanied by hardness growth of surface layer.

10.1   INTRODUCTION

Electroexplosive alloying (EEA) [1] and electron-beam processing (EBP) [2] are modern and effective methods of modification of the structure, phase composition and properties of metals and alloys surface. The instruments of this effect on the surface are pulse multiphase plasma jets and electronic beams accordingly.

Pulse multiphase plasma jets, used for the electroexplosive alloying, and low-energy high-current electron beams combine well with each other, having comparable values of the absorbed power density (about 10^5 W/cm^2), the square of irradiation surface (up to 10–15 cm^2) and the depth of hardening zone (about several tens of micrometers). Time of the impulse in the EEA is 100 μs, the EBP – 50÷200 μs.

The main idea of combined processing, including EEA and EBP, is in the leveling of the surface topography of processing and modification of the structure, phase composition and properties of alloying area.

The air of this work is the analysis of the structure and the properties of surface layers of steel 45 (C ≤ 0.45 wt. %), subjected to electroexplosive copper plating and subsequent electron-beam processing.

10.2   MATERIALS AND METHODS OF RESEARCH

EEA of the surface of the samples was carried out by electric explosion of copper foils with a thickness of 20 μm. Conditions for the implementation of the pulse of liquid-phase alloying is given by the quantity of charging voltage of the drive energy accelerator, the diameter of the charmer nozzle and the distance from it up to the sample, which amounted appropriately ~2.3 kV, 20 mm and 20 mm, correspondingly. With these parameters the depth and the radius of the zone of alloying were maximal. The processing time was 100 μs, absorbed power density on the axis of the jet was ~5.5 GW/m^2, dynamic pressure of shock-compressed layer near the surface was 11.2 MPa, the area of the surface of alloying was ~3 cm^2. The thickness of the area of alloying in the center of the area was ~25 μm. Its distinguishing feature is a strong influence on the results of the processing of the pressure jets on the surface, leading to a radial flow of the melt from the center of the area of the alloying to the periphery and even to the backlash. With this regime of the processing the maximum depth zone of the alloying, the degree of saturation of the alloying elements and the level of generated properties are achieved.

A pulsed electron-beam processing of the samples surface was carried out at the installation "SOLO." We used two variants of surface processing of steel electroexplosive alloying by copper. In the first case, we fixed the values of the energy density (E_S = 20 J/cm^2), the frequency (f = 0.3 Hz) and pulse duration (τ = 50 μs) and varied the number of pulses within n = 5...50. In the second case, the duration (τ = 50 μs), the frequency (f = 0.3 Hz) and the number of pulses (10 pls.) were fixed; electron beam energy density was varied within the E_S = 15...30 J/cm^2.

The research of the structure of the radiation surface, of the etched metallographic section (direct and seating metallographic section) of the modified samples were carried out by the methods of electron scanning and transmission electron microscopy, X-ray structure analysis. The change mechanical properties of a material were characterized by microhardness. The accuracy of the measurement is amounted to 7%. For phase identification diffraction analysis with the use of darkfield method and subsequent indicating of microelectronograms was is used.

10.3   RESULTS AND DISCUSSION

Electroexplosive alloying, transforming the structural-phase state of the surface layer of the processed material exercises, influence on the mechanical properties (Fig. 10.1.) The dependence of microhardness of steel, subjected to electroexplosive copper plating on the distance to the surface of the treatment has nonlinear character. This allows to identify the surface layer of thickness of about ~5 μm, microhardness of which is lower than that of microhardness of hardened steel; intermediate layer of thickness of about ~7 μm, which is marked on the Fig. 10.1. by vertical straight lines B and C, microhardness of which is higher or equal to the microhardness of hardened steel, and the transition layer of thickness of about ~30 μm, microhardness of which gradually falls to the quantity of the initial condition. Microhardness of the intermediate layer changes on the curve with a maximum, located at a depth of ~7 μm.

Microhardness of steel 45 after the EBP has the maximum value on the surface of radiation, and monotonically decreases with the depth. The thickness of the hardened layer after EBP is approximately 5 μm (Fig. 10.1, vertical line B), which, obviously, is determined by the selected steel processing regime.

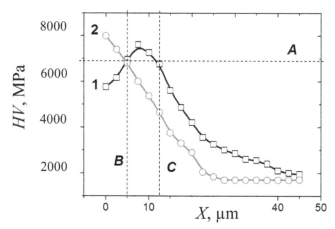

FIGURE 10.1   The profile of the steel 45 microhardness.

1-after electroexplosive copper alloying; 2 – after electron-beam processing according to the regime: 12 J/cm²; 50 μs; 0.3 Hz, 3 pls. A horizontal line **A** marks microhardness of steel 45, steel quenched from furnace heating (850°C, 1.5 h.)

The maximum of microhardness of steel after EBP exceeds the maximum values of microhardness of steel, hardened after furnace heating and after EEA. Electron-microscopic studies have shown that this is due to the formation of ultra fine-grained (0.54 ± 0.20 μm) structure on the surface of processing. The dimensions of martensite crystals in such grains change considerably within the following limits: cross – 30…50 nm, longitudinal – 120…500 nm.

The interpretation of the nonlinear dependence of microhardness on the distance to the surface of steel 45 treatment after electroexplosive copper plating follows from the analysis of the results presented in the work [1, 3]. The surface layer of steel 45, having the microhardness below the microhardness of steel 45 after furnace quenching is formed by the structure of the mesh crystallization of the melt, enriched by the atoms of copper, carbon and oxygen. The intermediate layer, the values of microhardness of which exceed the microhardness of steel 45 after furnace hardening, obviously, has formed as a result of the high-speed quenching of iron. A higher value of the microhardness of the layer (relative to the quenched steel) can be connected both with increased concentration of carbon and the existence of atoms of copper in this layer, so and with dispersion of the structure of the surface layer of steel due to the speed quenching caused by the impulse action. The rise of the microhardness of the hardened layer on the distance surface treatment can mean the reduction of volume fraction of residual austenite, stabilized by the atoms of copper and carbon. The subsequent hardness decrease is caused by the decrease of carbon concentration in the material, about that the change of morphology of martensite identified in the research armco-iron testifies: the transition from lamellar martensite, characteristic for carbon steel, to the packet martensite typical for low – and medium-carbon steel.

Electroexplosive alloying leads to the formation on the surface of the processed material a thin-layer coating, formed mainly droplet fraction of exploding wire. Subsequent EBP, without changing the elemental composition of the material, allows to fulfill high-speed homogenization of surface layer by high-intensity thermal effect.

It is established that the melting of the surface layer of the sample is fixing under the energy density of electron beam $Es \sim 15$ J/cm². This leads, on the one hand, to the removal of microcraters and the influxes of copper, forming a coating, on the other hand, to the formation of numerous drops of copper of spherical shape, the dimensions of which can range from 1 to 12 μm. The latter indicates the coagulation of copper coating located on the steel surface. It should be noted, that this EBP regime does not lead to the full smoothing of the surface of alloying – in some places of the sample the islands of copper remain.

Surface treatment EEA by electron beam with the energy density of the beam electrons 20...30 J/cm² is accompanied by widespread melting of the surface layer of steel – drops and islands of copper are not observed.

High speed crystallization of the melt leads to the formation of a dendrite structure. It is established, that the structure of the dendrites depends on the energy density of electron beam. When processing with the energy density of the electron beams 15...20 J/cm² mainly a dendritic structure with the axes of the first order forms (on the surface of radiation is so-called structure of cellular crystallization); dendrites with greater energy density have the axis of the first and the second order. It is obvious that the dendrite structure is determined by the speed of the cooling of the melt. It is shown [4, 5] that the axis of the second order are not formed already during the cooling rate, exceeding ~10^6 K/s. With the further increase of the cooling rate the complete degeneration of the dendritic growth and stabilization flat crystallization front is observed.

The increase of the energy density is accompanied not only by the change of morphology of dendrite structure, but also by increase in medium-sized dendrites. The estimates show that the dendrites of minimum medium size are formed under the processing of the steel surface by electron-beam with the energy density Es = 15 J/cm². The increase of the energy density from 15 up to 30 J/cm² is accompanied by the growth of dendrites of average sizes from 0.16 to 0.45 μm, that is, in ~3 times (Fig. 10.2, curve 1). The revealed facts allow us to conclude that the increase of the energy density of the beam electrons in the interval from 15 to 30 J/cm² leads to the decrease in the rate of cooling of the steel surface layer.

The average size of grains of the surface layer of steel depends on the rate of cooling. However, such a dependence, as for the elements of the dendrite structure, is not observed. As it follows from the analysis of the results presented in Fig. 10.2, curve 2, the average size of grains is increasing in 1,4 times in the revised interval of energy density of electron beam.

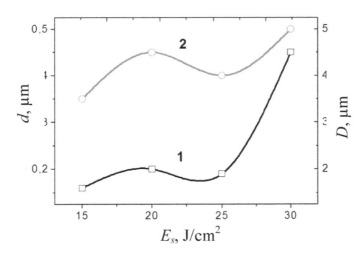

FIGURE 10.2   The dependence of the medium-sized dendrites d (curve 1) and grains D (curve 2) on the energy density of electron beam.

In Ref. [4], such a circumstance is explained by the fact that the size of the grains in the crystallized layer depends not only on the cooling rate (value of supercooling), but also on the number of active centers of grain nucleation in the melt.

Electron-beam processing of steel is accompanied by the formation of the microcracks on the surface. The reason is the thermal stresses, which are formed in the surface layer of the material due to high cooling rates. When the energy density of electron beam Es is ~15 J/cm² the cracks are located chaotically, their number is insignificantly. At the large values of Es the cracks break the surface of the specimen on the fragments, the average sizes of which vary within the range of 45...50 µm and practically do not depend on the energy density of electron beam. The depth of microcracks depends on the value of energy density of electron beam.

Figure 10.3 shows a diagram demonstrating the change of copper concentration in the surface layer of steel 45, subjected to EEA and following EBP. From the analysis of the results it follows that in the surface layer of thickness 4–5 µm (thickness layer of steel, subjected to analysis) the average concentration of copper is reduced from ~14 wt. % at a energy density of electron beam $Es = 15$ J/cm² up to 5.6 wt. % when $Es = 30$ J/cm². It should also be noted that on the surface of steel, processed by electron beam when Es = 15 J/cm^2, there are drops and islets, the concentration of copper in which can reach 100 wt. %.

The high speed crystallization of steel, alloyed by copper, and the following cooling do not always lead to the hardening of the surface layer. Hardness of the surface layer of steel, not treated by electron beam and treated by electron beam with the energy density of electron beam Es = 15 J/cm^2 is slightly lower than the hardness of steel 45, quenched in the water with the furnace heating and significantly below the hardness of steel, processed by electron-beam with the energy density of electron beam Es = 20…30 J/cm^2. Comparing the results, presented in Figs. 10.3. and 10.4, it can be found the connection between the concentration of copper in the surface layer of steel, and the value microhardness.

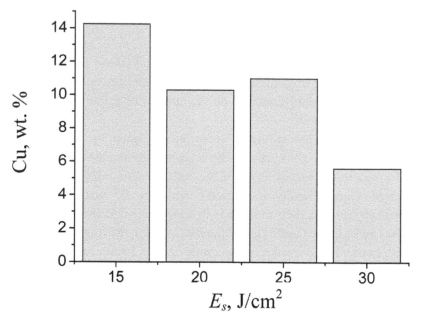

FIGURE 10.3   Change in the copper concentration in the surface layer of steel 45, subjected to electoexplosive alloying and subsequent electron-beam treatment with varying of electron beam energy density (50 µs, 0.3 Hz, 10 pls.).

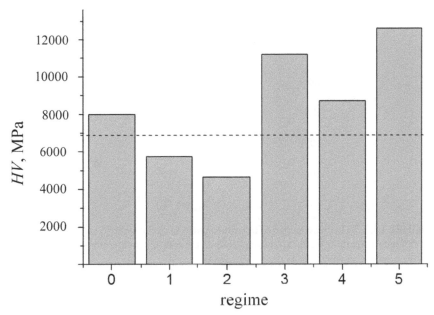

FIGURE 10.4 Microhardness of the surface layer of steel 45, subjected to the different types of effect: 0 – electron-beam treatment (12 J/cm², 50 µs, 0.3 Hz, 3 pls.); 1 – electroexplosive copper alloying; 2–5 – electroexplosive copper alloying and subsequent electron-beam processing (N = 10 pls., τ = 50 µs) at E_s = 15 (2), 20 (3), 25 (4), 30 (5) J/cm².

A horizontal line denotes a microhardness of steel 45, quenched from furnace heating (850° C, 1.5 h.)

Namely, the high values of the concentration of copper correspond to relatively low values of microhardness of a surface layer.

Functional dependence, connecting the concentration of copper in the surface layer and microhardness of the surface radiation is represented in Fig. 10.5.

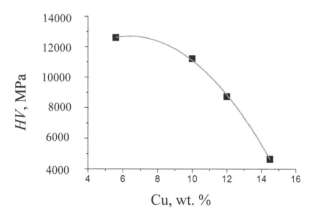

FIGURE 10.5   The dependence of the microhardness of the steel 45 surface, subjected to combined treatment (electroexplosive alloying. and subsequent electron-beam processing) on the copper concentration in the surface layer.

It is clearly seen that the microhardness of the surface layer of steel decreases with the increase of copper concentration. However, the linear correlation between these characteristics is not detected, which may denote indirect (by changing the parameters of the structure and phase composition) influence copper atoms on the hardness of the investigated steel, formed in the conditions of high-energy effect.

As it is shown above, the processing of the alloying surface by the electron beam with energy density of the electron beam 20 J/cm² and above is accompanied by extensive melting of the surface layer of steel. After 5...15 pulses of the electron beam effect the islands and the nodules of copper, presenting on the surface of the steel, subjected to EEA are not detected by methods of scanning microscopy. The surface of the samples is fully smoothed. After 25 and 50 pulses of the electron beam effect on the surface one can see a large number of craters.

It is found that the structure of the dendrites depends on the number of pulses of the electron beam effect. When the number of pulses being, changed within the limits of 5...15, on the surface of the steel the dendritic structure with the axes of the first order is formed (so-called structure of cellular crystallization). With a larger number of pulses of the electron beam (25 and 50 pls.) the dendrites mainly have the arises of the first and the second order.

The increase of the number pulses of the electron beam on the steel 45 surface leads to a decrease of cooling rate. The increase of the number pulses practically has no influence on the average size of the dendrites (Fig. 10.6, curve 1) and leads to a small increase in average size of the grains (Fig. 10.6, curve 2).

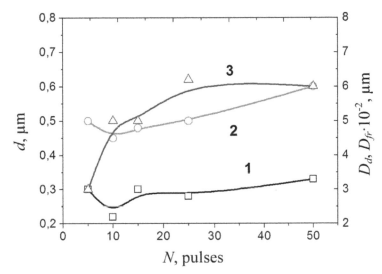

FIGURE 10.6   The dependence of dimensions of the medium-sized dendrites d (curve 1), grains D_d (curve 2) and fragments of D_{fn} (curve 3) on the number of pulses of electron beam.

Electron-beam treatment of steel is accompanied by the fragmentation of the surface. The average sizes of the fragments change in the range of 30...60 μm and increase with the number pulses of electron beam (Fig. 10.6, curve 3). This fact confirms the mentioned above assumption of the speed cooling decreasing with the increasing of radiation pulses number. In spite of the fact that the increase in the number of pulses of the electron beam leads to a decrease in a linear density of microcracks (growth of medium-sized fragments), their depth, judging by the size of the disclosure of microcracks, apparently, is increasing.

With the increasing of pulses number the copper concentration in the surface layer thickness of 4...5 μm is increasing steadily from ~8% at 5 pulses to 18% at 50 pulses, that is, more than in 2 times. It can be assumed, that one of the reasons of the revealed concentration of copper in the surface layer of steel is a displacement of copper atoms from the surface of sample volume with its multiple melting. In metallurgy of steel this process is named refining (clearing) of melts from harmful or unwanted elements (electron-beam remelting) [4].

The high speed crystallization of steel, alloyed by copper, and following cooling are accompanied by a significant increase of the surface layer hardness only in small pulses number of the electron beam (5 and 10 pulses). A further increase of the radiation pulses number is accompanied by a significant decrease in hardness of the surface layer of steel.

One can establish the relationship between the copper concentration in the surface layer of steel, and the quantity of microhardness. Namely, to the high values of the copper concentration correspond the low quantity of microhardness of the surface layer. However, the correlation between the characteristics of the steel is negligible, that may indicate the indirect (through the changing of the structure parameters and the phase of composition) effect of copper on the hardness of the surface layer of steel 45.

10.4   CONCLUSIONS

1. Electroexplosive copper plating of steel 45 is accompanied by the saturation of the surface layer of atoms of copper, carbon and oxygen. Subsequent high-speed cooling of steel is accompanied by the separation of the liquid phase and the formation of a surface layer with the structure of the cellular crystallization. The thickness of cellular structure is about 5 μm. The thickness of the layer of hardened steel 45, located at a depth of 5 μm, is about 7 μm.

2. Electron-beam treatment of steel is accompanied by the formation of microcracks dividing the surface of the specimen into the fragments. The average size of the fragments varies within the range of 30...60 μm and grows with an increase in the number of pulses of electron beam action.

3. The copper concentration in the surface layer thickness 4...5 μm increases from 8 wt. % at 5 pulses to 18 wt. % at 50 pulses, that is, more than in 2 times with the increase of pulses number of electron beam.

4. High speed crystallization of steel, alloyed by copper, and following after that cooling are accompanied by a significant (more than in ~1.5 times in competence with the hardness of steel 45, hardened from furnace heating) increase of the surface layer hardness only under small quantities of pulses (5 and 10 pulses).

KEYWORDS

- **copper**
- **electroexplosive alloying**
- **electron beam processing**
- **phase composition**
- **properties**
- **structure**

REFERENCES

1. The physical basis of electroexplosive alloying of metals and alloys. Bagautdinov A.Ya., Budovskikh Ye.A., Ivanov Yu. F, Gromov V.Ye. – Novokusnetsk: SibSIU, Publ. House 2007. 301 p.
2. Ivanov, Yu. F., Koval, N. N. Low-energy electron beams submillisecond duration: production and some aspects of application in the field of materials science. Ch.13 in the book "The structure and properties of advanced metallic materials." edited A.I. Potekayev. – Tomsk: Publ. House NTL, 2007. P. 345–382
3. Grigoriev S.V., Koval N.N., Ivanov Yu.F., et al. Electron beam modification of the surface of steel and hard alloys. "Plasma emission electronics." – Proceedings of II International кkreyndelevsky seminar. – Ulan-Ude, 2006. P. 113–120.
4. Miroshnichenko, I. S. Hardening from the liquid state. M.: Metallurgy, 1982. – 168 p.
5. Bigeyev, A. M. Metallurgy of the steel. Theory and technology of melting steel. M.: Metallurgy, 1977. 440 p.

PART IV

**INVESTIGATION METHODS
DEVELOPMENT FOR NANOSYSTEMS AND
NANOSTRUCTURED MATERIALS**

CHAPTER 11

THE STRUCTURAL ANALYSIS OF NANOCOMPOSITES POLYMER/ ORGANOCLAY FLAME-RESISTANCE

I. V. DOLBIN, G. V. KOZLOV, and G. E. ZAIKOV

CONTENTS

ABSTRACT

The structural analysis of nanocomposites polymer/organoclay flame-resistance was performed within the framework of percolation and multifractal models. The possibility of flame-resistance characteristics prediction on the basis of the indicated approach has been shown.

11.1 INTRODUCTION

At present polymeric materials are used widely as facing and heat-insulating coatings in buildings of different intendings, including social ones as well, that makes increased demands in their combustibility and flame-resistance with the purpose of safety ensuring. As it is known [1, 2], the introduction of organoclay in polymers, that is, nanocomposites polymer/organoclay formation, improves essentially the indicated above properties. At present several conceptions, developed for the description of this effect existed [1]. However, the indicated conceptions, even if they take into account nanocomposites structure and its influence on these materials flame-resistance, then only on purely qualitative level. In the present communication the percolation and multifractal models, developed for the description of nanocomposites polymer/organoclay structure and gas transport processes [3], will be used.

11.2 EXPERIMENTAL

The data for nanocomposites polymer/organoclay on the basis of polyamide-6 (PA-6), polyamide-12 (PA-12), polystyrene (PS) and polypropylene (PP), which are listed in table 11.1, were used for the relationships structure-flame-resistance characteristics. The maximum rate of heat release ϑ_{max}, measured with the use of a cone calorimeter according to the standards ASTM 1354–92 and ISO/DIS 13927 [2], the values of which are also listed in Table 11.1, was used as flame-resistance characteristic of the indicated nanomaterials.

TABLE 11.1 Characteristics of nanocomposites polymer/organoclay.

Matrix polymer	W_n, mass %	Organoclay type	J_{max}, kw/m²	j_n	j_n+j_{if}
PA-6	2	Exfoliated	686	0.030	0.087
	5	Exfoliated	378	0.075	0.218
PA-12	2	Exfoliated	1060	0.030	0.087
PS	3	Intercalated	1186	0.045	0.088
	3	Exfoliated	567	0.045	0.131
PP	2	Intercalated	450	0.030	0.176

11.3 RESULTS AND DISCUSSION

The authors [2] considered some aspects of nanofiller (organoclay) structure influence on nanocomposites flame-resistance. Firstly, the nanocomposite with exfoliated organoclay possesses higher flame-resistance, than the one with intercalated organoclay. Secondly, nanocomposites polymer/organoclay flame-resistance is higher than corresponding microcomposites at the same organoclay mass contents W_n. And thirdly, W_n increase results to the flame-resistance enhancement of the same nanocomposite.

These qualitative effects can be described quantitatively within the framework of percolation model of reinforcement and multifractal model of gas transport processes for nanocomposites polymer/organoclay [3, 4]. It has been supposed that two structural components are created for a barrier effect to fire spreading: actually organoclay and densely packed regions on its surface with relative volume fractions φ_n and φ_{if}, respectively. In other words, it has been supposed, that the value ϑ_{max} should be a diminishing function of the sum $(\varphi_n+\varphi_{if})$. For this supposition verification let us estimate the values φ_n and φ_{if}. The value φ_n is determined according to the well-known equation [5]:

$$\phi_n = \frac{W_n}{\rho_n},$$ (1)

where ρ_n is nanofiller density, which is estimated as follows [4]:

$$\rho_n = 188\left(D_p\right)^{1/3}, \text{kg/m3}, \tag{2}$$

where D_p is nanoparticle diameter, which is given in nanometers. For organoclay the arithmetical mean of three main sizes of its platelet is accepted as D_p.

Further the sum $(\varphi_n + \varphi_{if})$ can be calculated according to the equations [4]:

$$\phi_n + \phi_{if} = 1.955\phi_n b_\alpha \tag{3}$$

for the intercalated organoclay and

$$\phi_n + \phi_{if} = 2.191\phi_n b_\alpha \tag{4}$$

for the exfoliated one, where b_α is the parameter, characterizing the level of interfacial adhesion polymer matrix-nanofiller.

For further calculations as the first approximation $b_\alpha=1$ was accepted, that corresponds to perfect adhesion by Kerner [4]. In Fig. 11.1, the dependence of heat release maximum rate ϑ_{max} on the sum of nanofiller (organoclay) and interfacial regions relative volume contents is adduced, calculated according to the Eqs. (3) and (4) for intercalated and exfoliated organoclay, respectively. As one can see, all data correspond to one curve, which extrapolates to $\vartheta_{max} \approx 1450$ kw/m^2 at $(\varphi_n + \varphi_{if})=0$, that is, for unfilled polymer. This ϑ_{max} value corresponds to the indicated parameter mean value for the four matrix polymers, mentioned above. The sole essential deviation from the obtained curve is the data for nanocomposite on the basis of PP. This deviation can be due to the condition $b=1$ for all considered nanocomposites. The value b_α can be estimated more precisely with the aid of the following percolation relationship [4]:

$$\frac{E_n}{E_m} = 1 + 11\left(\phi_n + \phi_{if}b_\alpha\right)^{1.7}, \tag{5}$$

where E_n and E_m are elasticity moduli of nanocomposite and matrix polymer, respectively, the ratio of which gives the reinforcement degree of nanocomposite.

The data of Ref. [6] for four series of nanocomposites PP/Na⁺-mont-morillonite used for b_α calculation according to the Eq. (5), that gives average value $b \gg 2.87$. This value b_α application in the Eq. (3) demonstrated, that in that case the data for the considered nanocomposite corresponded to the common curve $\vartheta_{max}(\varphi_n + \varphi_{if})$ (Fig. 11.1).

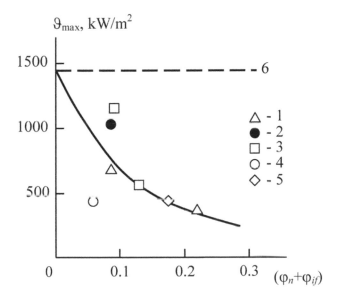

FIGURE 11.1 The dependence of heat release maximum rate ϑ_{max} on sum $(\varphi_n + \varphi_{if})$ for nanocomposites based on PA-6 (1), PA-12 (2), PS (3) and PP at $b_\alpha = 1.0$ (4) and $b_\alpha = 2.87$ (5). $6 - \vartheta_{max}$ average value for four matrix polymer.

Hence, the data of Fig. 11.1 suppose, that nanofiller together with densely packed interfacial regions creates the barrier effect, increasing flame-resistance of nanocomposite in comparison with matrix polymer. Let us consider the indicated effect physical significance. The calculation of the key parameter, heat release maximum rate ϑ_{max}, is based on the oxygen consummation principle. According to this principle, the heat, escaped at material combustion, is proportional to oxygen amount, requiring for its combustion [2]. Proceeding from this principle, the value ϑ_{max} should be connected with gas transport process characteristics for polymer nanocomposites. Nanocomposite permeability to gas coefficient P_n reduction in comparison with similar parameter for matrix polymer P_m within the framework of multifractal model is given as following [3]:

$$\frac{P_n}{P_m} = \left(\frac{\alpha_{am,n}}{\alpha_{am,m}}\right)^{d_m},$$

(6)

where $\alpha_{am,\,n}$ and $\alpha_{am,\,m}$ are relative fractions of amorphous phase for nanocomposite and matrix polymer, respectively, d_m is diameter of gas-penetrant (in our case – O_2) molecule.

Further theoretical values ϑ_{max} (ϑ_{max}^T) can be estimated according to a simple formula:

$$\vartheta_{max}^T = \vartheta_{max}^{av}\frac{P_n}{P_m},$$

(7)

where ϑ_{max}^{av} is an average value of heat release maximum rate for basic polymers ($\vartheta_{max}^{av} \approx 1450$ kw/m², Fig. 11.1). The values $\alpha_{am,\,m}$ are accepted according to the data [4] and the magnitudes $\alpha_{am,\,n}$ were estimated according to the equation [3]:

$$\alpha_{am,n} = \alpha_{am,m} - \left(\phi_n + \phi_{if}\right).$$

(8)

In fig. 11.2, the comparison of the values ϑ_{max} and ϑ_{max}^T is adduced for the studied nanocomposites, from which one can see, that even such simplest estimation gives a good correspondence of these parameters (the average discrepancy between ϑ_{max} and ϑ_{max}^T makes up less than 6%).

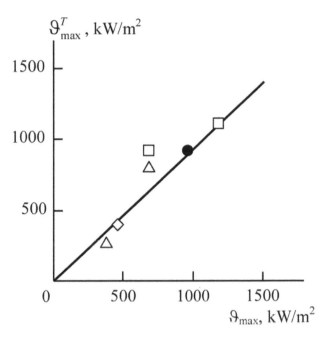

FIGURE 11.2 The comparison of experimental ϑ_{max} and calculated according to the Eq. (7) heat release maximum rate values. Conditional signs are the same that in Fig. 11.1.

11.4 CONCLUSIONS

Hence, the proposed approach allows one quantitative structural analysis of nanocomposites polymer/organoclay flame-resistance.

KEYWORDS

- **flame-resistance**
- **gas permeability**
- **nanocomposite**
- **organoclay**
- **structure**

REFERENCES

1. Lomakin, S. M., Zaikov, G. E. *Vysokomolek. Soed. B* (in rus.), **47**(1), 104–120 (2005).
2. Lomakin, S. M., Dubnikova, I. A., Berezina, S. M., Zaikov, G. E. *Vysokomolek. Soed. A* (in rus.), **48**(1), 90–105 (2006).
3. Kozlov, G. V., Mikitaev, A. K. *Structure and Properties of Nanocomposites Polymer/ Organoclay*, LAP LAMBERT Academic Publishing GmbH, Saarbrücken, p. 318 (2013).
4. Mikitaev, A. K., Kozlov, G. V., Zaikov, G. E. *Polymer Nanocomposites: Variety of Structural Forms and Applications*, Nova Science Publishers, Inc., New York, p. 319 (2008).
5. Sheng, N., Boyce, M. C., Parks, D. M., Rutledge, G. C., Abes, J. I., Cohen, R. E. *Polymer*, **45**(2), 487–506 (2004).
6. Antipov, E. M., Barannikov, A. A., Gerasin, V. A., Schklyaruk, B. F., Tsamalashvili, L. A., Fischer, H. R., Razumovskaya, I. V. *Vysokomolek. Soed. A* (in rus.), **45**(11), 1885–1899 (2003).

CHAPTER 12

THE METAL/CARBON NANOCOMPOSITES INFLUENCE MECHANISMS ON MEDIA AND ON COMPOSITIONS

V. I. KODOLOV and V. V. TRINEEVA

CONTENTS

ABSTRACT

The interaction mechanisms of liquid components or solvents with Metal/ Carbon Nanocomposites are considered. The activity of Metal/Carbon Nanocomposites in the different media is changed depending on the media polarity and dielectric penetration. IR and Raman spectra of finely dispersed suspensions of Metal/Carbon Nanocomposites demonstrate the change in the intensity in comparison with the pure medium. In IR spectra of suspensions the significant changes in the absorption intensity, especially in the regions of wave numbers close to the corresponding Nanocomposites oscillations, are observed. It is found that the effects of Nanocomposite influence on liquid media decreases with time and the activity of corresponding suspensions drops. Therefore the estimation of suspension stability (activity) as well as the suspension activation by ultrasound actions is proposed. At the same time the dependence of suspensions properties on the super small quantities of Nanocomposites is determined.

12.1 INTRODUCTION

The modification of materials by nanostructures including metal/carbon nanocomposites consists in the conducting of the following stages:
1. The choice of liquid phase for the making of finely dispersed suspensions intended for the definite material or composition.
2. The making of finely dispersed suspension with sufficient stability for the definite composition.
3. The development of conditions of the finely dispersed suspension introduction into composition.

At the choice of liquid phase for the making of finely dispersed suspension it should be taken the properties of nanostructures (nanocomposites) as well as liquid phase (polarity, dielectric penetration, viscosity).

It is perspective if the liquid phase enters completely into the structure of material formed during the composition hardening process.

12.2   INVESTIGATION OF FINELY DISPERSED SUSPENSIONS OF METAL/CARBON NANOCOMPOSITES

Information about the activity of Metal/Carbon Nanocomposites and about the synthesis of these Nanostructures is given in references [1, 2]. When the correspondent solvent, on the base of which the suspension is obtained, is evaporated, the re-coordination of nanocomposite on other components takes place and the effectiveness of action of nanostructures on composition is decreased [3].

The stability of finely dispersed suspension is determined on the optical density. The time of the suspension optical density conservation defines the stability of suspension. The activity of suspension is found on the bands intensity changes by means of IR and Raman spectra. The intensity increasing testify to transfer of nanostructure surface energy vibration part on the molecules of medium or composition. The line spreading in spectra testify to the growth of electron action of nanocomposites with medium molecules. Last fact is confirmed by X-ray photoelectron investigations.

The changes of character of distribution on nanoparticles sizes take place depending on the nature of nanocomposites, dielectric penetration and polarity of liquid phase.

Below characteristics of finely dispersed suspensions of metal/carbon nanocomposites are given (Figs. 12.1–12.4).

The distribution of nanoparticles in water, alcohol and water-alcohol suspensions prepared based on the above technique are determined with the help of laser analyzer. In Figs. 12.1 and 12.2, one can see distributions of copper/carbon nanocomposite in the media different polarity and dielectric penetration.

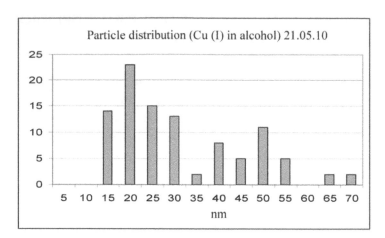

FIGURE 12.1 Distribution of copper/carbon nanocomposites in alcohol.

When comparing the figures we can see that ultrasound dispergation of one and the same nanocomposite in media different by polarity results in the changes of distribution of its particles. In water solution the average size of Cu/C nanocomposite equals 20 nm, and in alcohol medium – greater by 5 nm.

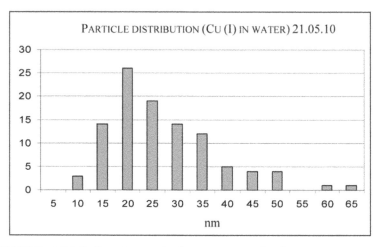

FIGURE 12.2 Distribution of copper/carbon nanocomposites in water.

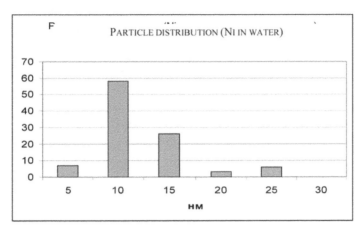

FIGURE 12.3   Distribution of nickel/carbon nanocomposites in water.

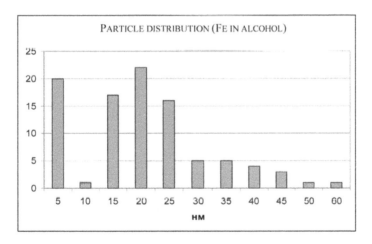

FIGURE 12.4   Distribution of iron/carbon nanocomposites.

Assuming that the nanocomposites obtained can be considered as oscillators transferring their oscillations onto the medium molecules, we can determine to what extent the IR spectrum of liquid medium will change, for example, polyethylene polyamine applied as a hardener in some polymeric compositions, when we introduce small and supersmall quantities of nanocomposite into the corresponding compositions [4].

IR spectra demonstrate the change in the intensity at the introduction of metal/carbon nanocomposite in comparison with the pure medium (IR spectra are given in Fig. 12.5). The intensities of IR absorption bands are directly connected with the polarization of chemical bonds at the change in their length, valence angles at the deformation oscillations, that is, at the change in molecule normal coordinates.

When nanostructures are introduced into media, we observe the changes in the area and intensity of bands, that indicates the coordination interactions and influence of nanostructures onto the medium (Figs. 12.5 and 12.6).

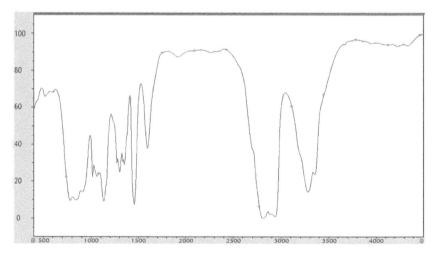

FIGURE 12.5   IR spectrum of polyethylene polyamine.

Special attention in PEPA spectrum should be paid to the peak at 1598 cm-¹ attributed to deformation oscillations of N-H bond, where hydrogen can participate in different coordination and exchange reactions.

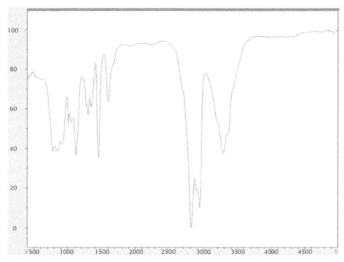

FIGURE 12.6. IR spectrum of copper/carbon nanocomposite finely dispersed suspension in polyethylene polyamine medium (ω (NC) = 1%).

In the spectra wave numbers characteristic for symmetric v_s(NH2) 3352 cm^{-1} and asymmetric v_{as}(NH$_2$) 3280 cm^{-1} oscillations of amine groups are present [5, 6]. There is a number of wave numbers attributed to symmetric v_s(CH$_2$) 2933 cm^{-1} and asymmetric valence v_{as}(CH$_2$) 2803 cm^{-1}, deformation wagging oscillations v_d(CH$_2$) 1349 cm^{-1} of methylene groups, deformation oscillations of NH v_d (NH) 1596 cm^{-1} and NH$_2$ v_d(NH$_2$) 1456 cm^{-1} amine groups. The oscillations of skeleton bonds at v(CN) 1059–1273 cm^{-1} and v(CC) 837 cm^{-1} are the most vivid. The analysis of intensities of IR spectra of PEPA and fine suspensions of metal/carbon nanocomposites based on it revealed a significant change in the intensities of amine groups of dispersion medium (for v_s(NH$_2$) in 1.26 times, and for v_{as}(NH$_2$) in approximately 50 times) (Fig.12.7).

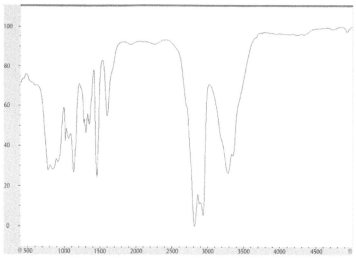

FIGURE 12.7 IR spectrum of nickel/carbon nanocomposite fine suspension in polyethylene polyamine (ω (NC) = 1%).

Such demonstrations are presumably connected with the distribution of the influence of nanoparticle oscillations onto the medium with further structuring and stabilizing of the system. Under the influence of nanoparticle the medium changes which is confirmed by the results of IR spectroscopy (change in the intensity of absorption bands in IR region). Density, dielectric penetration, viscosity of the medium are the determining parameters for obtaining fine suspension with uniform distribution of particles in the volume. At the same time, the structuring rate and consequently the stabilization of the system directly depend on the distribution by particle sizes in suspension. At the wide range of particle distribution by sizes, the oscillation frequency of particles different by size can significantly differ, in this connection, the distortion in the influence transfer of nanoparticle system onto the medium is possible (change in the medium from the part of some particles can be balanced by the other). At the narrow range of nanoparticle distribution by sizes the system structuring and stabilization are possible. With further adjustment of the components such processes will positively influence the processes of structuring and self-organization of final composite system determining physical-mechanical characteristics of hardened or hard composite system.

The effects of the influence of nanostructures at their interaction into liquid medium depend on the type of nanostructures, their content in the medium and medium nature. Depending on the material modified, fine suspensions of nanostructures based on different media are used. Water and water solutions of surface-active substances, plasticizers, foaming agents (when modifying foam concretes) are applied as such media to modify silicate, gypsum, cement and concrete compositions. To modify epoxy compounds and glues based on epoxy resins the media based on polyethylene polyamine, isomethyltetrahydrophthalic anhydride, toluene and alcohol-acetone solutions are applied. To modify polycarbonates and derivatives of polymethyl methacrylate dichloroethane and dichloromethane media are used. To modify polyvinyl chloride compositions and compositions based on phenolformaldehyde and phenolrubber polymers alcohol or acetone-based media are applied. Fine suspensions of metal/carbon nanocomposites are produced using the above media for specific compositions. In IR spectra of all studied suspensions the significant change in the absorption intensity, especially in the regions of wave numbers close to the corresponding nanocomposite oscillations, is observed. At the same time, it is found that the effects of nanocomposite influence on liquid media (fine suspensions) decreases with time and the activity of the corresponding suspensions drops. The time period in which the appropriate activity of nanocomposites is kept changes in the interval 24 h – 1 month depending on the nanocomposite type and nature of the basic medium (liquid phase in which nanocomposites dispergate). For instance, IR spectroscopic investigation of fine suspension based on isomethyltetrahydrophthalic anhydride containing 0.001% of Cu/C nanocomposite indicates the decrease in the peak intensity, which sharply increased on the third day when nanocomposite was introduced (Fig. 12.8).

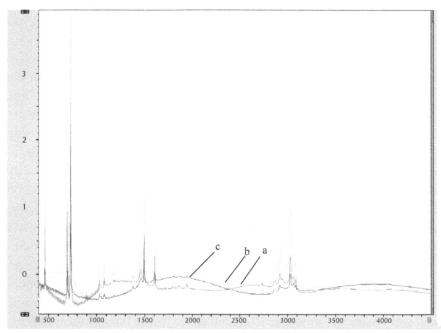

FIGURE 12.8 Changes in IR spectrum of copper/carbon nanocomposite fine suspension based on isomethyltetrahydrophthalic anhydride with time (a – IR spectrum on the first day after the nanocomposite was introduced, b – IR spectrum on the second day, c – IR spectrum on the third day).

Similar changes in IR spectra take place in water suspensions of metal/carbon nanocomposites based on water solutions of surface-active nanocomposites.

In Fig. 12.9, one can see IR spectrum of iron/carbon nanocomposite based on water solution of sodium lignosulfonate in comparison with IR spectrum of water solution of surface-active substance.

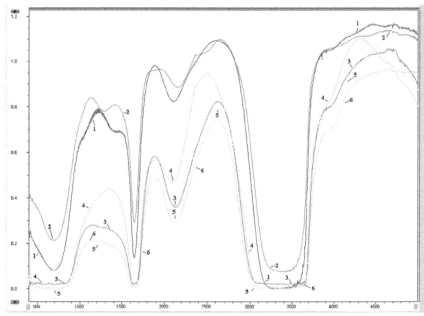

FIGURE 12.9 Comparison of IR spectra of water solution of sodium lignosulfonate (1) and fine suspension of iron/carbon nanocomposite (0.001%) based on this solution on the first day after nanocomposite introduction (2), on the third day (3), on the seventh day (4), 14th day (5) and 28th day (6).

As it is seen, when nanocomposite is introduced and undergoes ultrasound dispergation, the band intensity in the spectrum increases significantly. Also the shift of the bands in the regions 1100–1300 cm⁻¹, 2100–2200 cm⁻¹ is observed, which can indicate the interaction between sodium lignosulfonate and nanocomposite. However after two weeks the decrease in band intensity is seen. As the suspension stability evaluated by the optic density is 30 days, the nanocomposite activity is quite high in the period when IR spectra are taken. It can be expected that the effect of foam concrete modification with such suspension will be revealed if only 0.001% of nanocomposite is introduced.

12.3   GENERAL FUNDAMENTALS OF POLYMERIC MATERIALS MODIFICATION BY METAL/CARBON NANOCOMPOSITES

The material modification with the using of Metal/Carbon Nanocomposites is usually carried out by finely dispersed suspensions containing solvents or components of polymeric compositions. We realize the modification of the following materials: concrete foam, dense concrete, water glass, polyvinyl acetate, polyvinyl alcohol, polyvinyl chloride, polymethyl methacrylate, polycarbonate, epoxy resins, phenol-formaldehyde resins, reinforced plastics, glues, pastes, including current conducting polymeric materials and filled polymeric materials. Therefore it is necessary to use the different finely dispersed suspension for the modification of enumerated materials. The series of suspensions consist the suspensions on the basis of following liquids: water, ethanol, acetone, benzene, toluene, dichlorethane, methylene chloride, oleic acid, polyethylene polyamine, isomethyl tetra hydrophtalic anhydrite, water solutions surface-active substances or plasticizers. In some cases the solutions of correspondent polymers are applied for the making of the stable finely dispersed suspensions. The estimation of suspensions stability is given as the change of optical density during the definite time (Fig. 12.10).

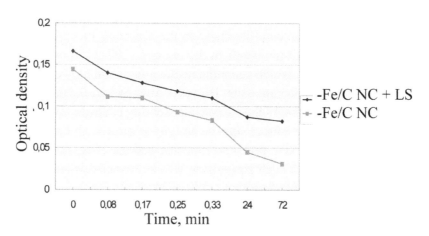

FIGURE 12.10   The change of optical density of typical suspension depending on time.

Relative change of free energy of coagulation process also may be as the estimation of suspension stability.

$$\Delta\Delta F = \lg k_{NC}/k_{NT} \qquad (1),$$

where $k_{NC}=\exp(-\Delta F_{NC}/RT)$ – the constant of coagulation rate of nanocomposite, $kNT=\exp(-\Delta F_{NT}/RT)$ – the constant of coagulation rate of carbon nanotube, ΔF_{NC}, ΔF_{NT} – the changes of free energies of corresponding systems (Fig. 12.11).

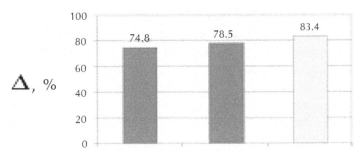

FIGURE 12.11   The comparison of finely dispersed suspensions stability.

The stability of metal/carbon nanocomposites suspension depends on the interactions of solvents with nanocomposites participation (Fig. 12.12).

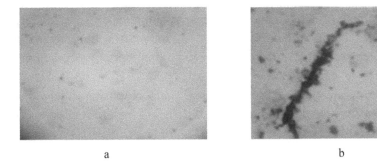

a   b

FIGURE 12.12   Microphotographs of Co/C nanocomposite suspension on the basis of mixture "dichlorethane – oleic acid" (a) and oleic acid (b).

The introduction of metal/carbon nanocomposites leads to the changes of kinematic and dynamic viscosity (Fig. 12.13).

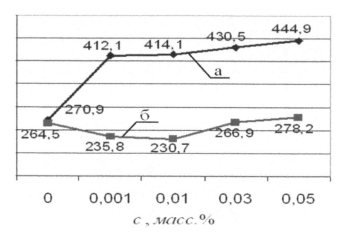

FIGURE 12.13   The dependence of dynamic and kinematic viscosity on the Cu/C nanocomposite quantity.

For the description of polymeric composition self organization the critical parameters (critical content of nanostructures critical time of process realization, critical energetic action) may be used.

The equation of nanostructure influence on medium is proposed –

$$W = n/N \exp\{an\tau^{\beta}/T\} \qquad (2)$$

where n – number of active nanostructures, n – number of nanostructure interaction, a – activity of nanostructure, τ – duration of self organization process, T – temperature, β – degree of freedom (number of process direction).

Thus, the application of Metal/Carbon Nanocomposites finely dispersed suspensions for the modification of different materials is perspective when the process theory is developed.

KEYWORDS

- **compositions**
- **interaction**
- **mechanisms**
- **media**
- **metal/carbon nanocomposites**
- **modifications**

REFERENCES

1. Kodolov, V. I., Khokhriakov, N. V. "Chemical physics of formation and transformation processes of nanostrcutures and nanosystems," Izhevsk State Agricultural Academy, Izhevsk, 2009, V.1 (365p), V.2 (415p).
2. Kodolov, V. I., Khokhriakov, N. V., Trineeva, V.V., Blagodatskikh, I. I. "Activity of nanostructures and its expression in nanoreactors of polymeric matrixes and active media," Chemical physics and mesoscopy, Vol.10, No. 4, 2008, pp. 448–460
3. Kodolov, V. I., Trineeva, V. V. "Perspectives of idea development about nanosystems self-organization in polymeric matrixes," In: A.M. Lipanov, Ed., The problems of nanochemistry for the creation of new materials, Institute for Engineering of Polymer Materials and Dyes, Torun. Poland, 2012, pp. 75–100.
4. Kodolov, V. I., Trineeva, V. V., Kovyazina, O. A., Vasilchenko, Yu. M. "Production and application of metal/carbon nanocomposites," In: A.M. Lipanov, Ed., The problems of nanochemistry for the creation of new materials, Institute for Engineering of Polymer Materials and Dyes, Torun. Poland, 2012, pp. 17–22.
5. "Spectral database for organic compounds AIST," 2001, http://riodb01.ibase.aist.go.jp/sdbs/cgi-bin/cre_index.cgi?lang=eng
6. Kazicina, L. A., Kupletskaya, N. B. "Application of UV-, IR- and NMR spectroscopy in organic chemistry," Chemistry Publ., Moscow, 1991.

CHAPTER 13

X-RAY STUDY OF THE INFLUENCE OF THE AMOUNT AND ACTIVITY OF CARBON METAL-CONTAINING NANOSTRUCTURES ON THE POLYMER MODIFICATION

I. N. SHABANOVA, V. I. KODOLOV, N. S. TEREBOVA, and YA. POLYOTOV

CONTENTS

ABSTRACT

In this chapter, the degree of the modification depending on the content of carbon metal-containing nanostructures is studied for polymers such as polycarbonate, polymethylmethacrylate and polyvinyl alcohol which have different structure and different content of oxygen bound to carbon.

It is shown that for obtaining the maximal degree of the modification of the above polymers, the content of carbon copper-containing nano-forms should be in the range of 10^{-2}–10^{-3}%. In this case the structure of the nano-modified polymers changes and becomes similar to the structure of the nanoforms. The degree of the modification depends on the amount of oxygen atoms in the polymer structure. The larger is the number of oxygen atoms, the smaller is the content of the nanoforms necessary for the polymer modification; in polycarbonate, the modification starts when the content of the nanostructures is 10^{-5}%, in polymethylmethacrylate, the modification is observed at the nanostructure content of 10^{-4}%, and for the polyvinyl alcohol modification the minimal content of 10^{-3}% is necessary.

The work was supported by the Program of Fundamental Research of the Ural Branch of the Russian Academy of Sciences, project № 12-Y-2–1034.

13.1   INTRODUCTION

High structure-forming activity of nanostructures allows their use for modification of materials. It is known that the modification of different materials with minute amounts of nanostructures improves material performance characteristics. At the same time, the mechanism of such influence of nanoforms on the structure and properties of materials has not been fully clarified yet. In the present chapter, explanations of the process are presented based on the XPS results using the modification of polymer systems as an example.

13.2   PROCEDURE

Carbon metal-containing nanostructures are multilayered nanotubes formed in a nanoreactor of a polymer matrix in the presence of the 3d-metal systems. Nano-carbon structures are prepared by an original procedure in the conditions of low temperatures (not higher than 400°C) which is pioneer investigations [1].

To provide uniform distribution of carbon metal-containing nanostructures throughout the bulk of the modified polymer is a very complicated task. The standard method for the uniform distribution of a nano-addition throughout the bulk of material is the preparation of fine-dispersed suspensions of nanoparticles in different media.

The method for the modification of polymer with carbon metal-containing nanostructures includes the preparation of fine-dispersed suspension (FDS) based on the solution of polymer in methylene chloride. Then, carbon metal-containing nanostructures are added into the prepared FDS. For the refinement and uniform distribution of carbon metal-containing nanostructures, the prepared mixture is treated with ultrasound. To prepare a film, the solvent is evaporated from the mixture at heating to 100°C.

13.3 EXPERIMENT

The XPS investigations were conducted on an X-ray electron magnetic spectrometer with the resolution 10^{-4}, luminosity 0.085% at the excitation by the AlKα line 1486.5 eV in vacuum 10^{-8}–10^{-10} Pa. In comparison with an electrostatic spectrometer, a magnetic spectrometer has a number of a advantages connected with the construction capabilities of X-ray electron magnetic spectrometers which are the constancy of luminosity and resolution independent of the energy of electrons, high contrast of spectra and the possibility of external actions on a sample during measurements [2].

The study of the variations of the C1s-spectrum shape lies in the basis of the investigations of the change of the polymer structure during nanomodification.

The identification of the C1s-spectra with the use of their satellite structure has been developed allowing the determination of the chemical bond of elements, the nearest surrounding of atoms and the type of sp-hybridization of the valence electrons of carbon in nanoclusters and materials modified with them. It is shown that in the C1s-spectrum, a satellite at the distance 22 eV from the main maximum is characteristic of the C-C–bond with the sp^2-hybridization of the valence electrons (graphite-like structure) and the relative intensity of the satellite is 10% of the main maximum. The satellite appears due to plasmon losses. When in the C1s-spectrum a satellite is observed at the distance 27 eV, the C-C-bonds are characteristic of sp^3-hybridization of the valence electrons of carbon atoms, that is, there is a diamond-like structure [3, 4]. In the C1s-spectrum of single-layer and multilayer nanotubes there are two satellites at 306 and 313 eV characteristic of C-C-bonds with sp^{2-} and sp^{3-} hybridization of the valence electrons on carbon atoms [5].

When there is a significant content of d-metal in nanomodified material, in addition to the C1s-spectrum, Me3s-spectra and valence band spectra are used. Based on the Van Vleck theory [6] a method is developed [7–9] describing the connection between the parameters of the multiplet splitting of the Me3s-spectra (the relation of the intensities of maxima and the distance between the multiplets) and the number of noncompensated d-electrons, atomic magnetic moment, localization of d-states in the surroundings of an own atom, d-p-hybridization of electrons of atoms in their nearest environment. Thus, the presence of changes in the Me3s-spectra shapes can provide the information about changes in the structure of nanomodified materials.

Since the surface of nanostructures has low reactivity, for increasing the nanostructure surface activity functionalization is used, that is, the attachment of certain atoms of sp- or d-elements to the atoms on the nanostructure surface and the formation of a covalent bond between them; the functionalization results in the formation of an interlink between the atoms of the nanostructure surface and the atoms of material.

The XPS study has shown [10] that the formation of the covalent bond between the atoms of the functional sp-groups and the atoms on the nanostructure surface is influenced by the electronegativity of the atoms of the components and the closeness of their covalent radii. Thus, it is most probable that the functionalization of nanostructures leads to the formation of the bond between phosphorus atoms and nanostructure d-metal atoms (Fe, Ni, Cu) and between nitrogen or fluorine atoms and carbon atoms of the nanostructures.

13.4   RESULTS AND DISCUSSION

The changes of the structure of the organic glass, polycarbonate and poly-vinyl alcohol modified with different amounts (10^{-5}, 10^{-4}, 10^{-3}, 10^{-2}, 10^{-1}%) of carbon copper-containing nanostructures have been studied.

Figure 13.1 presents the C1s-spectrum of the carbon copper-containing nanostructure, which consists of three components C-C (sp^2) – 284 eV, C-H – 285 eV, and C-C (sp^3)–286.2 eV. The presence of a small amount of the C-H component indicates an incomplete synthesis of nanostructures from the polymer matrix.

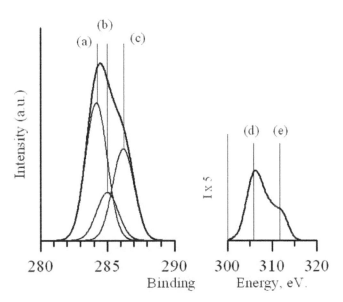

FIGURE 13.1   The XPS C1s-spectrum of carbon copper-containing nanostructures consisting of three components: a) C-C (sp^2) – 284 eV; b) C-H – 285 eV; c) C-C (sp^3) – 286.2 eV, and the satellite structure d) satellite (sp^2); e) satellite (sp^3).

The ratio of the maxima intensities of C-C (sp^2) and C-C (sp^3) depends on the dimension of the nanostructure: the larger is the surface area compared to the volume, the higher is the C1s-spectrum component with the sp^3 hybridization of valence electrons [3].

Figure 13.2 shows the C1s-spectra of polymethylmethacrylate (PMMA).

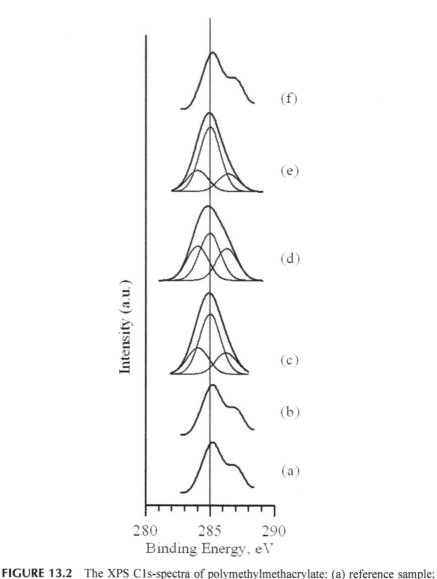

FIGURE 13.2   The XPS C1s-spectra of polymethylmethacrylate: (a) reference sample;
(b)nanomodified with carbon copper-containing nanostructures in the amount of 10^{-1}%;
(c) nanomodified with carbon copper-containing nanostructures in the amount of 10^{-2}%;
(d) nanomodified with carbon copper-containing nanostructures in the amount of 10^{-3}%;
(e) nanomodified with carbon copper-containing nanostructures in the amount of 10^{-4}%;
(f) nanomodified with carbon copper-containing nanostructures in the amount of 10^{-5}%.

In the reference sample, which is the film of organic glass, the bonds C-H and C-O (287 eV) prevail characterizing the organic glass structure. With increasing content of carbon copper-containing nanostructures in organic glass, starting from 10^{-4}%, the C1s-spectrum structure changes and the structure characteristic of carbon copper-containing nanostructure appears, namely, C-C (sp^2) and C-C (sp^3). The C-H component characterizes the remnants of the organic glass structure. Note that the C-O component is absent in the C1s-spectrum. During the organic glass modification, with increasing concentration of the nanostructure up to 10^{-3}%, in the C1s-spectrum the C-H component is decreasing and the C-C (sp^2) and C-C (sp^3) components characteristic of nanostructures are growing, that is, the degree of the polymer modification is growing. With a further increase in the nanostructure concentration (10^{-2}%), the degree of the polymer modification decreases and at the nanostructure content of 10^{-1}% the modification is absent. The structure of the C1s-spectrum becomes similar to that of the C1s-spectrum of the unmodified organic glass. Consequently, it is possible to judge about the degree of the polymer nanomodification by the ratio of C-C-bonds to C-H-bond in the C1s-spectrum. Maximal modification takes place at the nanostructure content of 10^{-3}%.

Similar results have been obtained for the nanomodification of polycarbonate (Fig. 13.3). The C1s-spectrum of the reference sample of polycarbonate containing a large amount of oxygen also consists of two components C-H (285 eV) and C-O (287 eV); however, the relative intensity of the C-O component is significantly larger than that observed for organic glass. When the content of nanostructures is in the range of 10^{-5}%–10^{-2}%, in the C1s-spectrum the structure characteristic of carbon copper-containing nanoform appears. The maximal change of the C1s-spectrum structure is observed when the nanostructure content in the polymer is 10^{-3}%. In contrast to organic glass, in polycarbonate the change of the structure starts at the nanoform content of 10^{-5}%. In this case, the C-O component in the C1s-spectrum decreases, which is indicated by the shift of the high-energy maximum from 287 eV to 286.2 eV in the C1s-spectrum; this corresponds to the binding energy of the C-C (sp^3) component. When the nanostructure content is in the range of 10^{-4}%–10^{-2}%, the C-O bonds are absent in the C1s-spectrum. In polycarbonate, at the nanostructure content of 10^{-1}%, no changes are observed in the polymer structure.

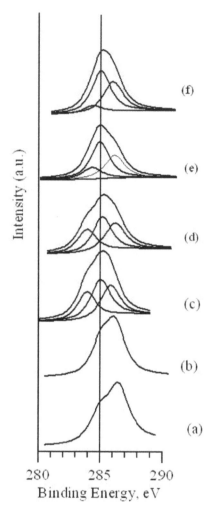

FIGURE 13.3 The XPS C1s-spectra of polycarbonate: (a) reference sample; (b) nanomodified with carbon copper-containing nanostructures in the amount of 10^{-1}%; (c) nanomodified with carbon copper-containing nanostructures in the amount of 10^{-2}%; (d) nanomodified with carbon copper-containing nanostructures in the amount of 10^{-3}%; (e) nanomodified with carbon copper-containing nanostructures in the amount of 10^{-4}%; (f) nanomodified with carbon copper-containing nanostructures in the amount of 10^{-5}%.

Figure 13.4 shows the C1s-spectra of polyvinyl alcohol (PVA) having in its structure the least content of oxygen in comparison with the other polymers under study. Similar to the reference samples of polycarbonate and organic glass, the C1s-spectrum of the reference sample of PVA (a PVA film) contains two components C-H and C-O; however, the C-O component is less intensive compared to C-H. The change of the PVA structure is observed only when it is modified with carbon copper-containing nanostructures, the content of which is in the range of $10^{-3}\%–10^{-2}\%$. When nanoforms are added into the PVA solution, in the C1s-spectrum structure the components appear which are characteristic of the C1s-spectrum of carbon copper-containing nanostructure (Fig. 13.1). When in PVA the nanostructure content is $10^{-1}\%$, no changes are observed in the polymer structure similar to the above-mentioned polymers; this can be explained by the processes of the nanostructure coagulation taking place when the nanostructure content in the polymer is high.

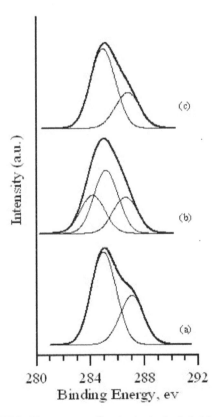

FIGURE 13.4 The XPS C1s-spectra of polyvinyl alcohol (a) reference sample; (b) nanomodified with carbon copper-containing nanostructures in the amount of $10^{-3}\%$; (c) nanomodified with carbon copper-containing nanostructures in the amount of $10^{-4}\%$.

A quite definite amount of the nanostructures is required for the modification of each of the studied polymers, that is, for changing their structure. The XPS studies show that the smaller is the number of oxygen atoms bound to carbon in the polymer, the larger is the amount of nanoparticles necessary for changing the polymer structure. For polycarbonate, it is 10^{-5}%, for organic glass – 10^{-4}%, and for polyvinyl alcohol – 10^{-3}%. When in the polymer the amount of nanoparticles is in the range from minimal up to 10^{-1}%, the C1s-spectrum changes; the component C-O (277.0 eV) disappears and the components C-C (sp^2) and C-C (sp^3) characteristic of a carbon metal-containing nanoform appear. The comparison of the structure of the studied polymers show high reactivity of the C-O bond in the CO_3 group in polycarbonate; the C-O bond reactivity decreases in the CO_2 group in organic glass, and its further decrease is observed in the CO group in polyvinyl alcohol. The disappearance of the component characterizing the C-O bond in the C1s-spectrum indicates the possibility of the replacement of this group of atoms by nanoparticles, that is, the formation of strong bonds between the polymer atoms and the atoms of the nanostructure surface. In this case, the polymer gains the structure-forming activity of nanostructures which are the centers of a new appearing structure. The breakage of the chemical bond of the C-O group with the nearest environment of the polymer atoms due to a different reactivity of the C-O bond takes place at different minimal contents of nanoparticles in the polymers.

Based on the results obtained it can be suggested that the larger is the content of oxygen atoms in the bond with carbon in the studied polymers, the larger is a change in the polymer structure and the larger is the formation of the regions in the polymer structure, which have a similar structure to that of carbon metal-containing nanoforms. The smaller is the number of oxygen atoms in the starting polymer, the larger is the amount of nanoforms necessary for the polymer structurization and the transformation of the polymer structure into the one similar to the nanoform structure during the modification.

The study was conducted on the influence of the growing activity of the interaction between the nanomodifier and the polymer resulting from the functionalization of the nanostructure surface with phosphorus atoms. The study of the functionalization mechanism [11] shows that the activity grows due to the formation of a covalent bond of P atoms and Cu atoms in the nanostructures because the Cu-P bond is stronger than the Cu-C bond. The nanostructure functionalization leads to an increase in the activity and, thus, to the increase of the degree of the material modification. It is seen from the comparison of the organic glass modified with nanostructures in the amount of 10^{-4}%; when the organic glass is modified with nonfunctionalized carbon copper-containing nanostructures, in the C1s-spectrum the ratio of the C-C (sp^2) component and the C-H component is 1:4 (Fig. 13.5a); and in the case of the modification of the organic glass with functionalized carbon copper-containing nanostructures, the ratio is 1:2 (Fig. 13.5b).

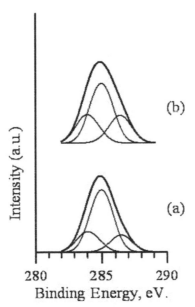

FIGURE 13.5 The XPS C1s-spectra of organic glass, (a) nanomodified with carbon copper-containing nanostructures in the amount of 10^{-4}%; (b) nanomodified with functionalized carbon copper-containing nanostructures in the amount of 10^{-4}%.

The experimental data show that the modification of the polymers occurs in the range from room temperature up to 100°C; at higher temperatures the modification is absent because the polymer decomposition starts.

13.4 CONCLUSION

In this chapter, it is shown that the degree of the nanostructure influence on the interaction with a polymer is determined by the content of nanostructures and their activity in this medium. The temperature growth blocks the development of self-organization in the medium. Thus, for describing the medium stucturization process under the nanostructure influence it is necessary to enter some critical parameters, namely, the content and activity of nanostructures and critical temperature.

The change of the polymer structure is accompanied by the change of their technological properties: the tensile strength of the studied films improves by 13%, the surface electrical resistance decreases by a factor of 3.3, and the transmission density of the films increases in the region close to an infrared one, which leads to an increase in the polymer heat capacity.

Thus, the X-ray studies show that

1. Mechanochemical and ultrasonic treatment facilitates the degree of the uniformity of the distribution of carbon metal-containing systems throughout the bulk of the modified medium and disintegration of groups into separate nanoparticles.

2. For obtaining the maximal degree of the modification of the studied polymers it is necessary that the content of carbon copper-containing nanoforms in them would be $\sim 10^{-3}$%. In this case, the structure of the nanomodified polymers changes and becomes similar to the structure of the nanoform.

3. The larger is the content of oxygen atoms bound to carbon atoms in the polymers, the larger is the degree of the polymer modification and the smaller is the percentage of nanoforms required for the beginning of modification.

4. High percentage of nanostructures can lead to the nanostructure coagulation and the absence of the polymer modification. In the studied polymers, it is observed at the nanostructure content of 10^{-1}%.

5. An increase in the activity of the interaction of the material and the nanostructures due to the nanostructure functionalization leads to an increase in the degree of the material structure modification.

KEYWORDS

- carbon copper-containing nanostructures
- functionalization
- modification
- polycarbonate
- polymethylmethacrylate (organic glass, PMMA)
- polyvinyl alcohol (PVA)
- satellite structure of C1s-spectra
- X-ray photoelectron spectroscopy (XPS)

REFERENCES

1. Kodolov, V. I., Khokhryakov, N. V., Chemical physics of the processes of the formation and transformations of nanostructures and nanosystems. Iz-vo IzhGSHA, Izhevsk, 2009. In two volumes. V. 1. 360 p. V. 2. 415 p.
2. Trapeznikov, V. A., Shabanova, I. N., Varganov, D. V., Dobysheva, L. V., et al. Izv. AN SSSR. Ser. fiz. 50. №9, 1677 (1986).
3. Makarova, L. G., Shabanova, I. N., Terebova, N. S., Zavodskaya laboratoria. Diagnostics of Materials. 71, №5, 26 (2005).
4. Makarova, L. G., Shabanova, I. N., Kodolov, V. I., A.P. Kuznetsov, J. Electr. Spectr. Rel. Phen. 137–140, 239 (2004).
5. Shabanova, I. N., Kodolov, V. I., Terebova, N. S., Trineeva, V. V., X-ray photoelectron spectroscopy in the investigation of carbon metal-containing nanosystems and nanostructured materials. Iz-vo "Udmurtski Universitet," Izhevsk. (2012). 250 p.
6. Auger- and X-ray photoelectron spectroscopy analysis of surface. edited by D. Briggs, M.P. Sikh. M.: Mir, 1987. P. 148.
7. Lomova, N. V., Shabanova, I. N., J. Electr. Spectr. and Rel. Phen. 2004. V.137–140. P. 511.
8. Shabanova, I. N., Terebova, N. S., Surf. and interface analysis. 2010. V. 46. P.846.
9. Shabanova, I. N., N.S. Terebova. Surface. X-ray, synchrotron and neutron investigations. 2012. №11. P1.
10. Shabanova, I. N., Terebova, N. S., Journal of Nanoscience and Nanotechnology Vol. 12, №11, 2012, pp. 8841–8844.

PART V

NANOSTRUCTURED MATERIALS PROPERTIES AND APPLICATION

CHAPTER 14

SEMICRYSTALLINE POLYMERS AS NATURAL HYBRID NANOCOMPOSITES: REINFORCEMENT DEGREE

G. M. MAGOMEDOV, K. S. DIBIROVA, G. V. KOZLOV, and G. E. ZAIKOV

CONTENTS

ABSTRACT

It has been shown that at semicrystalline polymers with devitrificated amorphous phase consideration as natural hybrid nanocomposites their anormalous high reinforcement degree is realized at the expense of crystallites partial recrystallization (mechanical disordering), that means crystalline phase participation in these polymers elastic properties formation. It is obvious, that the proposed mechanism is inapplicable for the description of polymer nanocomposites with inorganic nanofillers.

14.1 INTRODUCTION

As it is known [1], semicrystalline polymers, similar to widely applicable polyethylene and polypropylene, at temperatures of the order of room ones have devitrificated amorphous phase. This means that such phase elasticity modulus is small and makes up the value of the order of 10 MPa [2]. At the same time elasticity modulus of the indicated above polymers can reach values of ~1.0–1.4 GPa and is a comparable one with the corresponding parameter for amorphous glassy polymers. In case of the latters it has been shown [3, 4], that they can be considered as natural nanocomposites, in which local order domains (nanoclusters) serve as nanofiller and as loosely packed matrix of polymer within the framework of the cluster model of polymers amorphous state structure [5] is considered as matrix. In this case elasticity modulus of glassy loosely packed matrix makes up the value of order of 0.8 GPa and a corresponding parameter for polymer (e.g., polycarbonate or polyarylate) ~1.6 GPa. In other words, the reinforcement degree of loosely packed matrix by nanoclusters for amorphous glassy polymers is equal to ~2, whereas for the indicated above semicrystalline polymers it can exceed two orders. By analogy with amorphous [3, 4] and cross-linked [6] polymers semicrystalline polymers can be considered as natural hybrid nanocomposites, in which rubber-like matrix is reinforced by two kinds of nanofiller: nanoclusters (analog of disperse nanofiller with particles size of the order of ~1 nm [5]) and crystallites (analog of organo-clay with platelets size of the order of ~30–50 nm [7]). The clarification of abnormally high reinforcement degree mechanism allows to give an answer to the question, would this mechanism be applicable to polymer nanocomposites, filled with inorganic nanofiller (e.g., organoclay). Therefore, the purpose of this chapter is the study of reinforcement mechanism of rubber-like matrix of high density polyethylene (HDPE) at its consideration as natural hybrid nanocomposite.

14.2 EXPERIMENTAL

The gas-phase HDPE of industrial production of mark HDPE-276, GOST 16338–85 with average weight molecular mass 1.4×10^5 and crystallinity degree 0.723, measured by the sample density, was used.

The testing specimens were prepared by method of casting under pressure on a casting machine Test Samples Molding Apparate RR/TS MP of firm Ray-Ran (Taiwan) at material cylinder temperature 473 K, compression mold temperature 333 k and pressure of blockage 8 MPa.

The impact tests have been performed by using a pendulum impact machine on samples without a notch according to GOST 4746-80, type II, within the testing temperatures range T=213–333 K. Pendulum impact machine was equipped with a piezoelectric load sensor, that allows to determine elasticity modulus E and yield stress σ_Y in impact tests according to the techniques [8] and [9], respectively.

Uniaxial tension mechanical tests have been performed on the samples in the shape of two-sided spade with sizes according to GOST 11262–80. The tests have been conducted on a universal testing apparatus Gotech Testing Machine CT-TCS 2000, production of German Federal Republic, within the testing temperatures range T=293–363 k and strain rate of 2×10^{-3} s^{-1}.

14.3 RESULTS AND DISCUSSION

In Fig. 14.1, the temperature dependences of elasticity modulus E for the studied HDPE have been adduced. As one can see, at comparable testing temperatures E value in case of quasistatic tests is about twice smaller, than in impact ones. Let us note, that this distinction is not due to tests type. As it has been noted above, for HDPE with the same crystallinity degree at T=293 $k E$ value can reach 1252 MPa [10]. Let us consider the physical grounds of this discrepancy. The value of fractal dimension d_f of polymer structure, which is its main characteristic, can be determined by several methods application. The first from them uses the following equation [11]:

$$d_f = (d-1)(1+v),\tag{1}$$

where d is dimension of Euclidean space, in which a fractal is considered (it is obvious, that in our case d=3), v is Poisson's ratio, estimated according to the mechanical tests results with the aid of the equation [12]:

$$\frac{\sigma_Y}{E} = \frac{1-2v}{6(1+v)}.\tag{2}$$

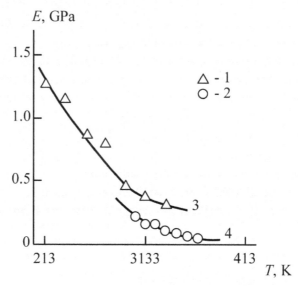

FIGURE 14.1   The dependences of elasticity modulus E on testing temperature T for HDPE, obtained in impact (1, 3) and quasistatic (2, 4) tests. 1, 2 – the experimental data; 3, 4 – calculation according to the equation (12).

The second method assumes the value d_f calculation according to the equation [5]:

$$d_f = d - 6.44 \times 10^{-10} \left(\frac{\phi_{cl}}{C_\infty S} \right)^{1/2}, \tag{3}$$

where ϕ_{cl} is relative fraction of local order domains (nanoclusters), C_∞ is characteristic ratio, S is cross-sectional area of macromolecule.

For HDPE $C_\infty = 7$ [13], $S=14.4$ Å^2 [14] and ϕ_{cl} value can be calculated according to the following percolation relationship [5]:

$$\phi_{cl} = 0.03 (1 - K)(T_m - T)^{0.55}, \tag{4}$$

where k is crystallinity degree, T_m and T are melting and testing temperatures, respectively. For HDPE $T_m \approx 400$ K [15].

And at last, for semicrystalline polymers d_f value can be evaluated as follows [16]:

$$d_f = 2 + K. \tag{5}$$

In Table 14.1 the comparison of d_f values, determined by the three indicated methods has been adduced (K change with temperature was estimated according to the data of Ref. [7]). As one can see, if in case of impact tests the calculation according to all three indicated methods gives coordinated results, then for quasistatic tests estimation according to the Eqs. (1) and (2) gives clearly understated d_f values, especially with appreciation of possible variation of this dimension for nonporous solids ($2 \leq d_f \leq 2.95$ [11]).

TABLE 14.1 The values of fractal dimension d_f of HDPE structure, calculated by different methods.

Tests type	T, K	d_f, the equation (1)	d_f, the equation (3)	d_f, the equation (5)
	293	2.302	2.800	2.723
	303	2.296	2.796	2.723
	313	2.272	2.802	2.713
Quasistatic	323	2.353	2.801	2.693
	333	2.248	2.799	2.673
	343	2.182	2.799	2.663
	353	2.170	2.800	2.643
	363	2.078	2.808	2.633

TABLE 14.1   *(Continued)*

Tests type	T, K	d_f, the equation (1)	d_f, the equation (3)	d_f, the equation (5)
	213	2.764	2.734	2.723
	233	2.762	2.741	2.723
	253	2.700	2.727	2.723
Impact	273	2.750	2.756	2.723
	293	2.680	2.729	2.723
	313	2.624	2.743	2.713
	333	2.646	2.766	2.763

Let us consider the causes of the indicated discrepancy in more details. At present, it has been assumed [1], that in case of semicrystalline polymers with devitrificated amorphous phase deformation in elasticity region, that is, at E value determination, the indicated amorphous phase is only deformed, that defines smaller values of both E and d_f. This conclusion is confirmed by disparity between d_f values, calculated on the basis of mechanical characteristics (the Eqs. (1) and (2)) and crystallinity degree (the Eq. (5)). And on the contrary, a good correspondence of d_f values, obtained by the three indicated methods (Table 14.1), assumes crystalline phase participance at HDPE deformation in elasticity region in case of impact tests (Fig. 14.1).

In Fig. 14.2, the temperature dependence of yield stress σ_Y has been adduced for HDPE in case of both types of tests.

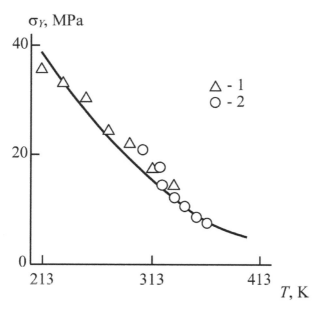

FIGURE 14.2 The dependence of yield stress σ_Y on testing temperature T for HDPE, obtained in impact (1) and quasistatic (2) tests.

As one can see, the data for quasistatic and impact tests are described by the same curve. As it has been shown in Ref. [5], σ_Y values are defined by contribution of both nanoclusters, that is, amorphous phase, and crystallites. Combined consideration of the plots of Figs. 14.1. and 14.2. demonstrates, that the distinction of d_f values, determined according to the Eqs. (1), (2) and (5), is only due to the indicated above structural distinction of HDPE deformation in an elasticity region.

The quantitative evaluation of crystalline regions contribution in HDPE elasticity can be performed within the framework of yield fractal conception [17], according to which the value of Poisson's ration in yield point v_Y can be estimated as follows:

$$v_Y = v\chi + 0.5(1-\chi),\qquad(6)$$

where v is Poisson's ratio in elastic strains region, determining according to the Eq. (2), χ is a relative fraction of elastically deformed polymer.

For amorphous glassy polymers it has been shown that χ value is equal to a relative fraction of loosely packed matrix $\varphi_{l.m.}$. In case of semicrystalline polymers in deformation process partial recrystallization (mechanical disordering) of a crystallites part can be realized, the relative fraction of which χ_{cr} is determined by the following equation [5]:

$$\chi_{cr} = \chi - (1 - K).$$ (7)

If to consider HDPE as natural hybrid nanocomposite, then amorphous phase (the indicated nanocomposite matrix) elasticity modulus E_{am} can be determined within the framework of high-elasticity conception, using the known equation [2]:

$$G_{am} = kNT,$$ (8)

where G_{am} is a shear modulus of amorphous phase, k is Boltzmann constant, n is a number of active chains of polymer network.

As it is known [5], in amorphous phase two types of macromolecular entanglements are present: traditional macromolecular "binary hookings" and entanglements, formed by nanoclusters, networks density of which is equal to v_e and v_{cl}, respectively. v_e value is determined within the framework of rubber high-elasticity conception [2]:

$$v_e = \frac{\rho_p N_A}{M_e},$$ (9)

where ρ_p is polymer density, N_A is Avogadro number, M_e is molecular weight of polymer chain part between macromolecular "binary hookings," which is equal to 1390 [18] and ρ_p value can be accepted equal to 960 kg/m³ [15].

In its turn, v_{cl} value can be determined according to the following equation [5]:

$$v_{cl} = \frac{\phi_{cl}}{C_\infty l_0 S},$$ (10)

where l_0 is length of the main chain skeletal bond, for HDPE equal to 1.54 Å [13].

E_{am} and G_{am} are connected by a simple fractal formula [11]:

$$E_{am} = d_f G_{am}.$$ (11)

Calculation according to the Eqs. (6) and (7) has shown that in case of HDPE quasistatic tests χ_{cr} value is close to zero and in case of impact tests χ_{cr}=0.400–0.146 within the range of T=231–333 K. Now reinforcement degree of HDPE, considered as hybrid nanocomposite, can be expressed as the ratio E/E_{am}. In Fig. 14.3, the dependence $E/E_{am}(\chi_{cr}^2)$ has been adduced (such form of dependence was chosen for its linearization). As one can see, this linear dependence, demonstrates E/E_{am} (or E) increase at χ_{cr} growth and is described analytically by the following relationship:

$$\frac{E}{E_{am}} = 590\chi_{cr}^2, \text{ MPa.}$$ (12)

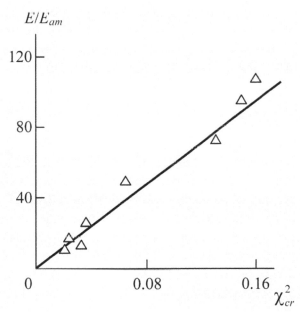

FIGURE 14.3 The dependence of reinforcement degree E/E_{am} on relative fraction of crystalline phase χ_{cr}, subjecting to partial recrystallization, for HDPE in impact tests.

The comparison of experimental E and calculated according to the Eq. (12) E^T elasticity modulus values for the studied HDPE has been adduced in Fig. 14.1. As one can see, the good correspondence between theory and experiment is obtained (the average discrepancy between E and E^T does not exceed 6%, that is, comparable with an error of elasticity modulus experimental determination.

Thus, the performed estimations demonstrated, that high values of reinforcement degree E/E_{am} for semicrystalline polymers, considered as hybrid nanocomposites (in case of studied HDPE E/E_{am} value is varied within the limits of 10–110) were due to recrystallization process (mechanical disordering of crystallites) in elastic deformation process and, as consequence, to contribution of crystalline regions in polymers elastic properties formation. It is obvious, that this mechanism does not work in case of inorganic nanofiller (e.g., organoclay). Besides, a nanofiller (crystallites) is formed spontaneously in a polymer crystallization process, that automatically cancels the problem of its dispersion at large k of the order of 70 mass %, whereas to obtain exfoliated organoclay at contents larger than 3 mass % is difficult [19]. Taking into consideration the indicated above factors it becomes clear, why nanocomposites polymer/organoclay maximum reinforcement degree does not exceed 4 [20].

14.4   CONCLUSIONS

The performed analysis has shown that at the consideration of semicrystalline polymers with devitrificated amorphous phase as natural hybrid nanocomposites their abnormally high reinforcement degree is realized at the expense of crystallites partial recrystallization (mechanical disordering), that means crystalline phase participation in the formation of these polymers elastic properties. It is obvious, that the proposed mechanism is inapplicable for the description of reinforcement of polymer nanocomposites with inorganic nanofiller.

KEYWORDS

- fractal dimension
- natural nanocomposite
- recrystallization
- reinforcement degree
- semicrystalline polymer

REFERENCES

1. Narisawa, I. *Strength of Polymeric Materials*. Chemistry Publishing House (in rus.), Moscow, p. 400 (1987).
2. Bartenev, G.M. and Frenkel, S.Ya. *Physics of Polymers*. Chemistry Publishing House (in rus.), Moscow, p. 432 (1990).
3. Kozlov, G.V. *Recent Patents on Chemical Engineering*, **4**(1), 53–77 (2011).
4. Kozlov, G.V. and Mikitaev, A.K. *Polymers as Natural Nanocomposites: Unrealized Potential*. Lambert Academic Publishing, Saarbrücken, p. 323 (2010).
5. Kozlov, G.V., Ovcharenko, E.N. and Mikitaev, A.K. *Structure of Polymer Amorphous State* (in rus.). RKhTU Publishing House, Moscow, p. 392 (2009).
6. Magomedov, G.M., Kozlov, G.V. and Zaikov, G.E. *Structure and Properties of Cross-Linked Polymers*. A Smithers Group Company, Shawbury, p. 492 (2011).
7. Tanabe, Y., Strobl, G.R. and Fisher, E.W. *Polymer*, **27**(8), 1147–1153 (1986).
8. Kozlov, G.V., Shetov, R.A. and Mikitaev, A.K. *Russian Polymer Science* (in rus.), **29**(5), 1109–1110 (1987).
9. Kozlov, G.V., Shetov, R.A. and Mikitaev, A.K. *Russian Polymer Science* (in rus.), **29**(9), 2012–2013 (1987).
10. Pegoretti, A., Dorigato, A. and Penati, A. *EXPRESS Polymer Lett.*, **1**(3), 123–131 (2007).
11. Balankin, A.S. *Synergetics of Deformable Body*. Ministry of Defence SSSR Publishing House, Moscow, p. 404 (1991).
12. Kozlov, G.V. and Sanditov, D.S. *Anharmonic Effects and Physical-Mechanical Properties of Polymers*. Nauka (Science) Publishing House (in rus.), Novosibirsk, p. 264 (1994).
13. Aharoni, S.M. *Macromolecules*, **16**(9), 1722–1728 (1983).
14. Aharoni, S.M. *Macromolecules*, **18**(12), 2624–2630 (1985).
15. Kalinchev, E.L. and Sakovtseva, M.B. *Properties and Processing of Thermoplastics* (in rus.). Chemistry Publishing House, Leningrad, p. 288 (1983).
16. Aloev, V.Z. and Kozlov, G.V. *Physics of Orientation Phenomena in Polymeric Materials* (in rus.). Polygraphservice and T, Nal'chik, p. 288 (2002).

17. Balankin, A.S. and Bugrimov, A.L. *Russian Polymer Science* (in rus.), **34**(5), 129–132 (1992).
18. Graessley, W.W. and Edwards, S.F. *Polymer*, **22**(10), 1329–1334 (1981).
19. Miktaev, A.K., Kozlov, G.V. and Zaikov, G.E. *Polymer Nanocomposites: Variety of Structural Forms and Applications*. Nova Science Publishers, Inc., New York, p. 319 (2008).
20. Liang, Z.-M., Yin, J., Wu, J.-H., Qiu, Z.-X. and He, F.-F. *Europ. Polymer J.*, **40**(2), 307–314 (2004).

CHAPTER 15

RESEARCH OF STRUCTURE AND POLISHING PROPERTIES OF NANOPOWDERS BASED ON CERIUM DIOXIDE

M. I. LEBEDEVA, E. L. DZIDZIGURI, and L. A. ARGATKINA

CONTENTS

ABSTRACT

The correlation between the polishing ability and the shape of particles was detected. It is established that adding fluorine to samples significantly affects the morphology and dimensional characteristics of REE oxides solid solution based on CeO_2. The hypothesis about the positive influence of high content of CeO_2 in a solid solution on the polishing ability of the material has not been confirmed.

15.1   INTRODUCTION

Polishing is an integral part of manufacture of optical glasses, jewels, mirrors, elements of electronic technics.

However, the factors determining the high polishing ability have still not been defined clearly. For example, it is considered [1] that polishing powders with a high content of CeO_2 have the best properties. Some studies claim that powders with a lamellar form of particles polish better; other studies insist that powders with a spherical form of particles have better polishing characteristics.

Materials obtained by the oxalate method show high polishing properties. However, the production based on this technology is not profitable. Polishing powders, synthesized by the carbonate method, have sufficiently low cost, but their polishing ability is low. To increase it, fluoridation is used. Comparison of the performance properties of nonfluorinated powders obtained by the carbonated technology and samples obtained via oxalates shows that the polishing ability of the latter is significantly – about two times – higher.

As a result, it becomes necessary to identify those properties of the powders, which form the polishing ability of a material, regardless of production conditions and composition.

15.2   EXPERIMENTAL

In this research the samples, received using different quantities of a precipitator from oxalates and carbonates, were studied. In addition, fluorine in amounts ranging from 0% to 15% was introduced into the powders, synthesized by the carbonate technology.

Na_2CO_3 was used as precipitant in obtaining a solid solution based on CeO_2 by carbonates precipitation from the chloride solution of rare earth elements (REE). Precipitator consumption was 80% and 95% based on the amount required for the stoichiometric reaction for the precipitation of average REE carbonate – $Ln (CO_3)_{1,5}$. Then, the samples were fluorinated in order to increase CeO_2 concentration in the solid solution, which, according to [1], should lead to the increase of polishing ability of the material. The cumulative equation is:

$$Ln_2(CO_3)_3 + 2NH_4F = 2Ln(CO_3)F + 2NH_3 + CO_2 + H_2O \tag{1}$$

The polish was obtained via oxalates using similar technology. The precipitator flow ($NH_4C_2O_4$) was varied: 92%, 110%, 130%, 160%, 200%.

The received powders were calculated at temperature 1050°C, thus there were transformations in the material according to schemes (2), (3) and (4):

$$Ln_2(CO_3)_3 = Ln_2O_3 + 3CO_2 \tag{2}$$

$$Ln(CO_3)F \rightarrow Ln_xO_y + LnOF + LnF_3 + CO_2 \tag{3}$$

$$Ln_2(C_2O_4)_3 \rightarrow Ln_xO_y + CO_2 + H_2O \tag{4}$$

The solid solution of oxides of rare earth elements in the natural mineral loparite has the following composition: Ce – 50 wt. %, La – 25 wt. %, Nb – 12 wt. %, Pr – 6 wt. %, the rest – Sm, Eu, Gd. The content of elements in synthesized powders may differ from the original due to the use of various amounts of precipitators or to lanthanide binding in the fluorine-containing compounds.

Morphology and dimensional characteristics of the polish were investigated by X-ray diffractometer "Rigaku" in Fe-Kα radiation and "Difrey" in Cr-Kα radiation. Approximation methods and Fourier analysis were used for the processing of experimental data. The average size of coherent-scattering regions (CSR) was calculated by two methods: Selyakov-Scherrer equation and Selivanov-Smyslov technique.

Log-normal law of crystalline particles size distribution is the basis of Selivanov-Smyslov technique. This technique can be used in the investigation of nanomaterials with a spherical shape and a unimodal distribution and with the size of structural components from 5 to 150 nm [2].

Determination of the lattice constant "a" was performed by extrapolation of the experimental data with the angle of 90° in coordinates "½ (cos2 θ / sin θ + cos2 θ / θ) – lattice period," where θ is the angle of diffraction.

Microimages were obtained using a scanning electron microscope "CamScan."

Polishing powder ability is the mass loss of K-8 glass in the form of a disk with the diameter of 75 mm and the thickness of 8 mm to 14 mm with a deviation of not more than 1.25 μm during 15 min of polishing process in the machine "6ShT-100 M" or any other machine that provides the required technological processing parameters with pressure on the polished surface of 180 g/cm^2.

Quantitative estimation of polishing ability was performed on the basis of mass loss of K-8 glass during 15 min of the sample polishing in the pad. For each glass sample two consecutive polishing ability measurements of one sample of the powder in the one pad were conducted. An arithmetic average of two consecutive polishing ability measurements was accepted for the measurement result.

15.3   RESULTS AND DISCUSSION

Phase X-ray diagrams of solid solution based on cerium dioxide, obtained by synthesis via carbonates using 95% of precipitator is presented in Fig. 15.1. The research has shown that samples with fluorine mass fraction less than 2% contain only one phase: cerium oxide, except for the sample containing 0.5% fluorine. Further introduction of fluorine from 3 to 8% leads to oxyfluoride phase (LnOF) separation. There are three phases (CeO_2, LnOF and LnF_3) when fluoride content is more than 9%. Samples with 80% of precipitator have the same phase composition. The results of the quantitative phase and structural analysis of samples, synthesized via carbonates, with different amounts of precipitator and with high polishing ability are shown in Table 15.1.

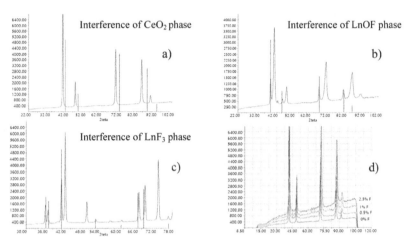

FIGURE 15.1 Diffractograms of solid solution based on cerium dioxide, synthesized via carbonates, with 95% of precipitator containing: a) 0% F, b) 5% F, c) 15% F, d) Overlaid diffractograms of nanopowders with 0%, 0.5%, 1% and 2.5% F.

TABLE 15.1 Results of X-ray-phase analysis for samples obtained with 80% and 95% of precipitator.

Conditions of production	Fluorine content, weight fraction, %	Phase composition	Number of phases, weight fraction, %			The lattice constant of CeO_2, nm	Polishing ability, mg/15 min
			CeO_2	LnOF	LnF$_3$		
Synthesis via carbonates with 95% of precipitator	0	CeO_2	100	—	—	0.5532	93
	2	CeO_2	100	—	—	0.5542	115
	7	CeO_2, CeOF	68	32	—	0.5467	118
	11	CeO_2, CeF_3, CeOF	60	16	24	0.5459	115
Synthesis via carbonates with 80% of precipitator	0	CeO_2	100	0	0	0.5519	65
	4	CeO_2, CeOF	60.4	39.6	0	—	105
	11	CeO_2, CeF_3, CeOF	41.1	29.5	29,4	0.5441	105

In the lot of nanopowders obtained at 80% of precipitator, two samples have a high polishing ability – those with 4% and 11% F. The sample with 4% F consists of CeO_2 and LnOF phases, the sample with 11% F consists of CeO_2, LnOF and LnF_3 phases. In the lot of nanopowders obtained at 95% of precipitator, the highest polishing ability is demonstrated by the sample with 7% F, which consists of two phases (CeO_2 and LnOF). The samples with 2% and 11% F in this lot have also shown a similar polishing ability. The first one consists only of the CeO_2 phase, and the second contains all three phases of CeO_2, LnOF and LnF_3.

Polishing nanopowders, synthesized via oxalates, have only one phase – solid solution of REE oxides. However, the polishing ability of these materials is high.

Thus, the phase composition of nanopowders obtained by the carbonate technology has all possible variants: there are samples that contain only one phase – CeO_2, two phases – CeO_2, LnOF, and all three phases – CeO_2, LnOF and LnF_3. However, materials with different phase compositions have the same polishing ability. Therefore, the phase composition is not a determining factor in the formation of the polish operational properties.

The size of the lattice constant of pure CeO_2 is 0.541 nm. The size analysis of the lattice constant of the solid solution based on REE oxides indicates that the concentration of CeO_2 in the samples obtained with 80% of precipitator is larger than the concentration of CeO_2 in the samples obtained with 95% of precipitator. The measurement results of the polishing ability show that nanopowders obtained with 95% precipitator have the highest polishing ability. Consequently, the hypothesis [1] that the increase in concentration of CeO_2 in the solid solution should lead to an increase in the material polishing ability, is not confirmed by experimental data.

The calculated values of the lattice constant of solid solution based on CeO_2, obtained via oxalates, are shown in Table 15.2 and Fig. 15.2. As seen in Fig. 15.2, all the extrapolation curves have a negative angle of inclination and are located significantly higher than the values of the lattice constant of pure CeO_2. Using a precipitant which is 10% more than stoichiometrically proved leads to an abrupt increase in the period of the crystal lattice of the solid solution based on CeO_2. The lattice constant decreases with further increase in the quantity of precipitant. At the same time the polishing ability varies nonmonotonically, but remains at a high level.

TABLE 15.2 Results of the structural analysis for samples obtained via oxalates.

Precipitator content, %	Lattice constant, nm	Polishing ability, mg/15 min
92	0.5506	120
110	0.5537	105
130	0.5534	110
160	0.5541	115
200	0.5514	100

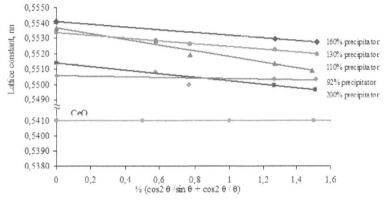

FIGURE 15.2 Change of the lattice constants of samples obtained via oxalate depending on the content of precipitator.

The lattices periods of the samples obtained via oxalates at 200% of precipitant and the samples synthesized via carbonates with 0% F at 80% of precipitator have similar values, but their polishing ability differs significantly (almost two times). Nanopowders with the highest operational properties (118 mg/15 min – a sample obtained by the carbonate technique with 7% F at 95% of precipitant and 120 mg/15 min – a sample obtained via oxalates with 92% of precipitant), have different periods of the lattices (0.5467 nm and 0.5506 nm, respectively). Therefore, there is no correlation between lattice constant and polishing ability of the polish.

The most representative microimages of the samples obtained via oxalates and carbonates with following fluorination are shown in Figs. 15.3 and 15.4, respectively. The nanopowder obtained via carbonates with 0%

fluorine represents the plates consisting of globular particles (Fig. 15.3a). The increase of fluorine content leads to loosening of the plates, formation of porous aggregates and their subsequent destruction, down to separate particles (Fig. 15.3b, c).

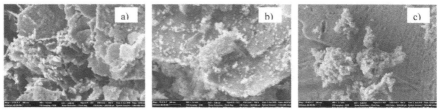

FIGURE 15.3 Microimages of the studied samples obtained with 95% precipitator containing: a) 0% F, b) 2% F, c) 7% F.

Nanopowders obtained via oxalates, consist of globular particles of 50–100 nm (Fig. 15.4).

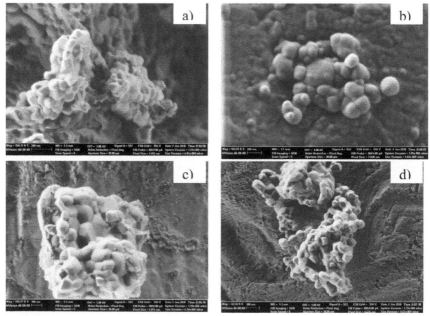

FIGURE 15.4 Microimages of the studied samples obtained via oxalates with precipitator: a) 92% b) 110% c) 130% d) 200%.

The analysis of all the studied nanopowders shows that the samples obtained via carbonates with 95% of precipitator and via oxalates with 92% of precipitant have the highest polishing ability (118–120 mg/15 min). Both of these materials are aggregates consisting of globular particles. Consequently, the morphology of nanopowders affects the polish operational properties significantly: powders with globular particles have a higher polishing ability.

Average particle sizes calculated by microimages are summarized in Table 15.3. With the addition of 0.5% F, a sharp decrease in particle size in the samples obtained via carbonates with 95% of precipitant is observed. Upon further fluorine addition up to 2.5–3.0%, the particle size increases. Particle sizes of solid solution based on cerium dioxide obtained via oxalates with 92% of precipitator, and carbonates with 0% F are in a good agreement. But the polishing ability of these materials is quite different.

Average sizes of CSR, calculated according to Selyakov-Scherrer equation and Selivanov-Smyslov technique (Table 15.3) are in good agreement.

A high polishing ability is observed both in nanopowders consisting of large globular particles of 50–100 nm (oxalate technology) and in samples consisting of particles of smaller size 20–30 nm (carbonate technology). Nanopowders' polishing ability and average sizes of CSR correlate in a similar way. Thus, the results imply that the operational properties do not depend on particle sizes and CSR of material.

TABLE 15.3 Results of structural analysis for the samples obtained via carbonates with following fluorination at 80 and 95% of precipitator.

Conditions of production	Varied factors	Particle shape	Average particle size, nm $D_{el.micr}$	Average CSR size, nm $D_{Sel-Sher}$	$D_{Smys-Sel}$	Polishing ability, mg/15 min
Synthesis via carbonates with 95% of precipitator	0% F	Plates consisting of globular particles,	50	22	18	93
	2.0% F	single globular particles	20	16	17	115
	7.0% F	Globular	30	—	—	118
	11.0% F	Globular	—	—	30	115

TABLE 15.3   *(Continued)*

Conditions of production	Varied factors	Particle shape	Average particle size, nm	Average CSR size, nm		Polishing ability, mg/15 min
			$D_{el.micr}$	$D_{Sel\text{-}Sher}$	$D_{Smys\text{-}Sel}$	
Synthesis via carbonates with 80% of precipitator	0% F	Platelike*	49	—	28	65
	4.0% F	Plates consisting of globular particles, single globular particles	15	19	16	105
	11.2% F	Globular	50	35	18	105
	92% of precipitator	Globular	50	40	36	120
	110% of precipitator	Globular	100	58	43	95
Synthesis via oxalates	130% of precipitator	Globular	80	40	—	110
	160% of precipitator	Globular	100	67	—	115
	200% of precipitator	Globular	100	53	45	100

*The magnification of the microimage is not sufficient for a more detailed description of morphology. Most likely the plates consist of globular particles.

Note: $D_{el.micr}$: Average CeO_2 particle sizes calculated by microimages.

$D_{Sel\text{-}Sher}$: Average CSR size calculated by Selyakov–Sherrer equation.

$D_{Smys\text{-}Sel}$: Average CSR size calculated by Selivanov–Smyslov equation.

15.4 CONCLUSIONS

The correlation between the polishing ability and the shape of particles was detected.

It is established that adding fluorine to samples significantly affects the morphology and dimensional characteristics of REE oxides solid solution based on CeO_2.

A clear connection between nanopowders' lattice constant and operational properties of the material was not found.

Dependence between dispersion characteristics (particle sizes and CSR) and polishing ability of nanopowders based on CeO_2 was not found.

The hypothesis about the positive influence of high content of CeO_2 in a solid solution on the polishing ability of the material has not been confirmed.

KEYWORDS

- **dimensional characteristics**
- **morphology**
- **nanopowders based on cerium dioxide**
- **polishing ability**
- **X-ray-phase analysis**

REFERENCES

1. G. S. Khodakov, N. L. Kudryavtseva, Physicochemical Processes of Polishing of Optical Glass, Mashinostroenie, Moscow, 1985 [in Russian].
2. E.L. Dzidziguri, E. N. Sidirova, Ultradispersed environment. X-ray diffraction methods for the study of nanomaterials, Ucheba, Moscow, 2007 [in Russian].

CHAPTER 16

TEMPERATURE FIELD CONTROL AND FORECAST IN NANOCOMPOSITIONAL MATERIALS

N. I. SIDNYAEV, Y. S. ILINA, and D. A. KRYLOV

CONTENTS

ABSTRACT

This chapter describes methods of temperature fields calculation in nano-compositional materials, which are considered as nanosystems with an appropriate medium. Mathematical models describing real physical processes are given. Calculate formulas for nanomaterials are offered.

16.1 INTRODUCTION

Study of the properties of materials in nano-state has a great importance for the development of fundamental science and for the practical application of such nanomaterials in devices of nano- and microsystem technology [1, 3]. The decrease of the characteristic dimensions of particles down to the values matching the value of the length of the de Broglie waves in a solid body leads to the quantization of energy levels and strong change in the particle polarizability. The increasing role of relaxation of surface atoms is accompanied by a change of the electronic structure of point defects. With the decrease of the particles diameter the ratio of the squares of their surfaces to their inner volume increases and the percent of the surface atoms grows. This leads to a change in the conditions of phase balance, to the reduction of the melting temperatures, to the change of the solubility limits, to the shift of the phonon spectrum to the region of short wavelength, to a change of catalytic properties, to nanophases initiation and other effects. It determines the principal opportunities for the creation of new nano-materials with unique physical and chemical properties [1–4]. Recently, the unique optical and magnetic properties of nanoparticles on the basis of rare-earth elements have made them the central point of different researches and development. The special properties of oxide compounds of rare-earth elements are largely responsible for the availability of vacant electron shells of the atom and the corresponding energy levels remain as a part of a solid structure being discrete.

Thanks to this feature nanoscale devices on the basis of doped with rare earth metals glasses were widespread, especially in the field of telecommunications as a part of the amplifiers of radiation in optical fibers. Getting oxides of rare metals in crystalline form involves certain difficulties due to the refractoriness of the substance, therefore, obtaining of complex multicomponent systems based on powders of initial oxides requires expensive equipment to achieve high temperatures up to 1.500°C. In nanomaterials with a hierarchical structure adsorption occurs simultaneously in the pores of all types up to the complete filling of micropores. In micropores the potential of adsorption is high due to the addition of dispersive potentials close the walls of the pores which leads to adsorption heat increasing and to the filling of these pores at low pressures. Then adsorption continues in meso- and macropores mechanisms of polymolecular adsorption, then capillary condensation begins in mesopores with continuing polymolecular adsorption in large meso- and macropores. After the limit completion of mesopores only polymolecular adsorption lasts on the surface of macropores.

In nanomaterials with a hierarchical structure of pores the phase transformation occurs simultaneously in the pores of all types up to the complete filling of micropores. In micropores potential adsorption increased due to the addition of dispersive potentials of the closely set pore walls which leads to increase of temperature and the filling of such time using nanotubes at a small relative pressures. Phase transformations can be managed in meso- and macropores mechanisms in the form of nanotubes (Fig. 16.1). After the limit completion of mesopores temperature fields can be controlled.

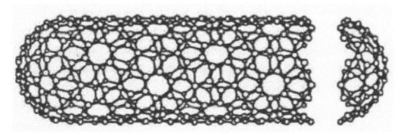

FIGURE 16.1 Nanotube structure.

Figure 16.2 presents the research of nanocomposites using atomic force microscopy [3].

For example, during the flow of practically irreversible polycondensation reactions fractal aggregates grew and the number of possible permutations between particles at fixing them on the skeleton of fractals reduced. This leads to the reduction of entropy of mixing which causes increase of the value of the energy of mixing (Gibbs energy change) [5, 6].

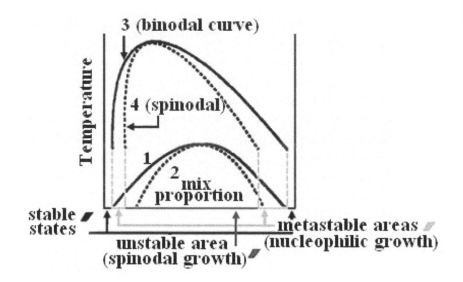

FIGURE 16.2   General view of the phase diagrams for the polymer system.

Lately nanotubes have been used in the cooling plants as capillary heat bends. In particular, researchers at the Purdue University have developed a water cooling system for heated electronics based on carbon nanotubes. Thanks to the use of nanotubes, represented as microscopic capillaries carrying water, this new cooling system does not need a pump that makes water circulate in the traditional systems of water cooling.

In the developed system (Fig. 16.3) water plays the role of a refrigerant, like Freon in refrigerating machines. Due to the small diameter of the capillaries, about 50 nm, which are carbon nanotubes, water, passing through them, evaporated completely removing a great amount of heat from the microchip (*see* Fig. 16.4). Excessive pressure of the steam, which then condenses in the refrigerator, constitutes the driving force, which makes water circulate throughout the cooling system [7–10].

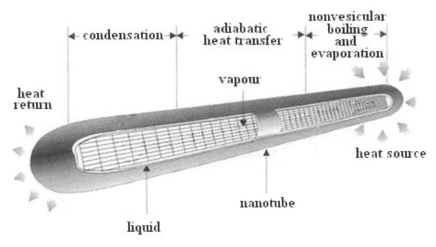

FIGURE 16.3   Movement of water in carbon nanotube leading to the removal of heat from the surface of the heated element.

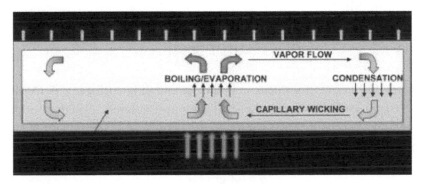

FIGURE 16.4   Scheme of cooling system based on carbon nanotubes.

This cooling system is not the first system developed by scientists of the Purdue University for cooling of heated electronics. Some time ago a system of air cooling was already developed, where surface of radiators was presented by a great amount of carbon nanotubes. Thanks to this the effective area of radiators increased several hundred times, allowing the system to discharge a large amount of heat into the air and, thus, increasing its effectiveness.

In this chapter, the object of research is presented by two-phase nano-structural environment, where temperature fields can pass the border of phase transition. These phase transitions are associated with allocation and absorption of large quantities of heat, which have a significant effect on temperature field in this environment and its dynamics.

Boundary conditions of three types can be set on the borders of the design volume. In addition, we will consider nanomaterials with nanotubes inside, which will be positioned as additional sources (or outflows) of heat.

16.2   BASIC EQUATIONS AND THE PROBLEM STATEMENT

16.2.1   BASIC EQUATIONS

The basic equation for the development of methods for forecasting of distribution of temperature fields is the equation of heat balance in the integral form [8]:

$$\frac{\partial}{\partial t}\iiint_V Hdv = -\iint_\Sigma \vec{q}\vec{n}\, d\sigma + \iiint_V Fdv. \tag{1}$$

where V is a control volume, Σ is the surrounding surface, T is for time, $H=H(u)$ is enthalpy (internal energy) per unit volume, $\vec{q}$ is the vector of heat flow, $\vec{n}$ is an external normal to the volume, F is the flow of heat per unit volume. In the soil the heat flux $\vec{q}$ is determined by the Fourier law:

$$\vec{q} = -\lambda\, grad\, u \tag{2}$$

λ is a conduction coefficient, u is for temperature.

Phase transitions and jumps of coefficients of thermal conductivity on the borders of different types of nanomaterials or structural elements generate gaps in the decisions. In the absence of such gaps on the basis of the equations (1) and (2) shows the equation of heat propagation in a differential form:

$$c_v \frac{\partial}{\partial t}u = \frac{\partial}{\partial x}(\lambda\frac{\partial}{\partial x}u) + \frac{\partial}{\partial y}(\lambda\frac{\partial}{\partial y}u) + \frac{\partial}{\partial z}(\lambda\frac{\partial}{\partial z}u) + F(x,y,z) \tag{3}$$

There $c_v = dH / du = \rho c$ – volumetric coefficient of thermal conductivity, c – specific heat, ρ – density. Here, it is believed that in case of presence of a phase transition, where enthalpy suffers a jump, volumetric coefficient of thermal conductivity has the form

$$c_v = \rho c + Q\delta(u - u^*),$$

where u^* is the temperature of the phase transition, Q is heat of the phase transition, $\delta(u - u^*)$ – delta-function.

Enthalpy in the general nonlinear case is a monotonically increasing function of temperature. In the simplest case, it can be considered

$$H = H_0 + c_l(u - u^*), \ u > u^*,$$

$$H = H_0 + c_f(u - u^*) - Q, \ u < u^*.$$

There c_l is a volumetric heat capacity of liquid phase, c_f is the volumetric heat capacity of solid phase, H_0 is the definition of enthalpy of. In the case $H_0 = 0$, the enthalpy equals zero when $u = u_*$.

In the case when the heat conductivity coefficient is constant Eq. (3) can lead to the equation of heat conductivity:

$$\frac{c_v}{\lambda}\frac{\partial}{\partial t}u = \frac{\partial^2}{\partial x^2}u + \frac{\partial^2}{\partial y^2}u + \frac{\partial^2}{\partial z^2}u + \frac{F(x, y, z)}{\lambda}. \tag{4}$$

In this model is assumed that when $u<0$ there can be a region partially containing material in the solid phase, where the percentage of particles of firm substance depends on the temperature, therefore enthalpy at $u<0$ is a nonlinear function of temperature. For a description of the areas with partial content of solid substances more complex two-phase models exist, where the ratio between liquid and solid phases depends upon the time from the start of transition to a solid phase. Such models require calculation of two equations of heat transfer. These models are not considered as long-term current processes are assumed to take place.

16.2.2   CALCULATION OF BREAKS

Two different analytical methods and two different types of numerical methods for solving problems with discontinuities are used. In the first method discontinuities are segregated and continuous regions are described by the Eqs. (1) or (2). Some boundary conditions derived from Eqs. (1) and (2) are put in areas of discontinuity. The phase transition boundary conditions are the following [8, 13]:

$$U(H_2 - H_1) = \lambda_1 \left(\frac{\partial}{\partial \bar{s}} u\right)_2 - \lambda_2 \left(\frac{\partial}{\partial \bar{s}} u\right)_1, \ u_1 = 0, \ u_2 = 0. \tag{5}$$

There U is a the speed of spreading of the break surface in the direction of its normal $\bar{s}$, indices 1 and 2 denote the corresponding values on different sides of the break, normal $\bar{s}$ directs from region 1 to region 2. It is assumed that there is an enthalpy jump when $u=0$ because of the phase transition heat. Also the leap of heat conductivity for solid and liquid substances is taken into account, but it is not necessary for the existing of a heat conductivity jump.

For stationary breaks associated with the jump of heat conductivity on the boundary of two different environments, boundary conditions are the following [9, 13, 14]:

$$\lambda_1 \left(\frac{\partial}{\partial \bar{s}} u\right)_1 - \lambda_2 \left(\frac{\partial}{\partial \bar{s}} u\right)_2 = 0, \ u_1 = u_2. \tag{6}$$

In the one-dimensional case when all depends upon the x-coordinate these equations take the form of:

$$U(H_1 - H_2) = \lambda_1 \left(\frac{\partial}{\partial x} u\right)_1 - \lambda_2 \left(\frac{\partial}{\partial x} u\right)_2, \ u_1 = 0, \ u_2 = 0;$$

$$\lambda_1 \left(\frac{\partial}{\partial x} u\right)_1 - \lambda_2 \left(\frac{\partial}{\partial x} u\right)_2, \ u_1 = u_2.$$

16.2.3 BOUNDARY CONDITIONS ON THE BOUNDARY OF THE DESIGN VOLUME AND THEIR APPLICATION FOR CALCULATION OF CONSTRUCTIONS

There are three types of boundary conditions on the border of the design volume. The boundary condition of the first type:

$$u = \mu(t)$$.   (7)

Implementation sample: a surface with given temperature. The boundary condition of the second type:

$$-\lambda \frac{\partial}{\partial n} u = v(t).$$ (8)

Application examples: flat thermostatic device, allowing to regulate the heat flow from the surface, when $v(t) = 0$ is a heat-insulating surface, standard boundary condition on the lateral border of the design volume. The boundary condition of the third type:

$$\lambda \frac{\partial}{\partial n} u = -\sigma(u - \mu(t))$$ (9)

It is used to describe the area of contact with double-sided heat-conducting surface, where temperature coincides with the temperature of nanomaterial on the one side, and on the other – there is a given temperature $\mu(t)$. This condition simulates a thin layer of the thermally conductive nanomaterial. The value of σ is calculated by the formula: $\sigma = \lambda_m / h$, λ_m is thermal conductivity of nanomaterial, h – its thickness. Application examples: nanoconstrucion that is subjected to the temperature drops from outside and inside.

16.3   DIFFERENCE SCHEMES FOR SOLVING OF THE HEAT CONDUCTION EQUATION

16.3.1   EXPLICIT DIFFERENCE SCHEMES FOR THE HEAT EQUATION

Let us consider the problem of heat conduction in a bar which initial temperature equals to zero [8]. Let the temperature of the left end be fixed, and on the right end there is a heat exchange with the environment, so that the heat flux is proportional to the temperature difference of the end of the bar and the environment (Fig. 16.5).

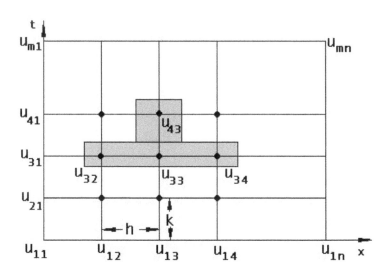

FIGURE 16.5   Difference mesh for the heat equation.

Let the temperature of nanomaterial be determined by the function $g(t)$. In other words, the problem to solve is the following:

$$u_t = u_{xx}, \ 0 < x < 1, \ 0 < t < \infty,$$
$$\begin{cases} u(0,t) = 1, \\ u_x(1,t) = -\left[u(1,t) - g(t)\right], \end{cases} \ 0 < t < \infty, \tag{10}$$
$$u(x,0) = 0, \ 0 \le x \le 1.$$

For solving this problem by the finite difference method [2] it is necessary to build a rectangular grid, which nodes are determined by formulas (Fig. 16.5)

$$x_j = jh, \; j = 0, 1, 2, ..., n,$$
$$y_i = ik, \; i = 0, 1, 2, ..., m.$$

It is important to note that the values u_{ij} on the left and the bottom sides of the grid on Fig. 16.2 are given by the initial and boundary conditions and our task is to find the rest of the values u_{ij}. To solve this problem, we replace the partial derivatives in the equation of heat conductivity by their finite-difference approximations

$$u_t = \frac{1}{k}\left[u(x, t+k) - u(x,t)\right] = \frac{1}{k}\left[u_{i+1,j} - u_{i,j}\right],$$
$$u_{xx} = \frac{1}{h^2}\left[u(x+h,t) - 2u(x,t) + u(x-h,t)\right] = \frac{1}{h^2}\left[u_{i,j+1} - 2u_{i,j} + u_{i,j-1}\right].$$

Let us substitute these expressions into the equation $u_t = u_x$ and solve the resulting equation attempting to find the values of the function at the top temporal layer. The result is

$$u_{i+1,j} = u_{i,j} + \frac{k}{h^2}\left[u_{i,j+1} - 2u_{i,j} + u_{i,j-1}\right]. \tag{11}$$

This is a required formula because it expresses the solution at a given moment of time by the solution in the previous moment of time (the index i refers to the time variable). Figure 16.4 highlights those values that are included in this formula.

Now the computing can be brought. But first it is necessary to approximate the derivative at boundary condition on the right end.

$$u_x(1,t) = -\left[u(1,t) - g(t)\right].$$

As a result of approximation we obtain

$$\frac{1}{h}\left[u_{i,n} - u_{i,n-1}\right] = -\left[u_{i,n} - g_i\right], \tag{12}$$

where values $g_i = g(ik)$ are known. Here we replaced $u_x(1,t)$ by the left differential derivative, since the right differential derivative would require the values of the function outside the grid.

From Eq. (12) we find

$$u_{i,n} = \frac{u_{i,n-1} - hg_i}{1+h}. \tag{13}$$

Let us start the calculation using Eqs. (11) and (13).

16.3.2 CALCULATION ALGORITHM OF AN EXPLICIT SCHEME

Step 1. Finding the solution on a grid layer $t = \Delta t$, using an explicit formula Fig. 16.6.

$$u_{2,j} = u_{1,j} + \frac{k}{h^2}\left[u_{1,j+1} - 2u_{1,j} + u_{1,j-1}\right], \; j = 2,3,...,n-1.$$

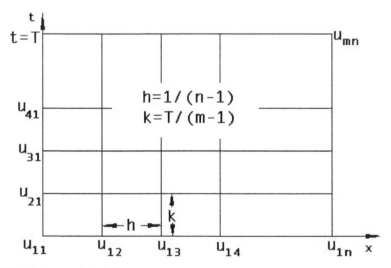

FIGURE 16.6 Explicit difference scheme: n – the number of mesh points along the axis x; m – the number of mesh points along the axis y; $h=1/(n-1)$, $k=T/(m-1)$.

Step 2. The value $u_{2,n}$ can be found by the Eq. (13)

$$u_{2,n} = \frac{u_{2,n-1} + hg_2}{1+h}.$$

After steps 1 and 2 we obtain the solution for $t = \Delta t$. To obtain a solution with $t = 2\Delta t$ (second to last line in Fig. 16.5) you can repeat steps 1 and 2, climbed up one line, that is, increasing i by 1 and using $u_{i,j}$ from the previous line. Similarly the decision is calculated in subsequent moments of time $t = 3\Delta t, 4\Delta t, \ldots$

16.3.3 COMMENTS

1. The explicit scheme has a serious drawback. If the time step is sufficiently large in comparison with the step of x, the rounding errors may become so large that the obtained solution is meaningless. The ratio of steps t and x depends on the equation and the boundary conditions, but in the general case the time step should be much smaller than the coordinate step. It is proved that for the applicability of the explicit scheme there must be $k / h^2 \leq 0.5$.

2. We have the following rule of thumb: if you reduce the steps Δt and Δx then the error of approximation by partial derivatives of finite differences will also decrease, but the smaller the net, the more calculations are necessary to take and, consequently, the greater will be the rounding error.

3. For hyperbolic tasks,

$$u_{tt} = u_{xx}, \ 0 < x < 1, \ 0 < t < \infty,$$
$$\begin{cases} u(0,t) = g_1(t), \\ u(1,t) = g_2(t), \end{cases} 0 < t < \infty,$$
$$\begin{cases} u(x,0) = \phi(x) \\ u_t(x,0) = \psi(x) \end{cases}, \ 0 \leq x \leq 1,$$

you can also build an explicit difference scheme. For this u_{tt} and u_{xx} must be approximated by central difference derivatives.

$$u_{tt} \cong \frac{1}{k^2}\left[u(x,t+k)-2u(x,t)+u(x,t-k)\right],$$

$$u_{xx} \cong \frac{1}{h^2}\left[u(x+h,t)-2u(x,t)+u(x-h,t)\right],$$

and the initial condition is

$$u_t(x,0) \cong \frac{1}{k}\left[u(x,k)-u(x,0)\right] = \frac{1}{k}\left[u(x,k)-\phi(x)\right].$$

As a result we obtain the following explicit scheme for calculation of the value of $u(x,t+k)$:

$$u(x,t+k) = 2u(x,t)-u(x,t-k)+\left(\frac{k}{h}\right)^2\left[u(x+h,t)-2u(x,t)+u(x-h,t)\right]. \quad (14)$$

From Eq. (14) it can be seen that for the calculation of the decision on the particular temporary layer you must know the solution at the previous two layers. Therefore, for the start of the countdown, you must use the initial condition of speed

$$\frac{1}{k}\left[u(x,k)-\phi(x)\right] = \psi(x),$$

from which can be obtained the following equation:

$$u(x,k) = \phi(x)+k\psi(x),$$

that is, the value of the solution at $t = \Delta t$. The decision in subsequent moments of time can be found with the explicit Eq. (14).

16.3.4   IMPLICIT SCHEME FOR THE HEAT EQUATION

Let us consider the following problem [2, 7],

$$u_t = u_{xx}, \ 0 < x < 1, \ 0 < t < \infty,$$

$$\begin{cases} u(0,t) = 0, \\ u_x(1,t) = 0, \end{cases} \quad 0 < t < \infty,$$

$$u(x,0) = 1, \qquad 0 \le x \le 1. \tag{15}$$

Let us use the following finite-difference approximations for partial derivatives u_t and u_{xx}:

$$u_t(x,t) = \frac{1}{k}[u(x,t+k) - u(x,t)],$$

$$u_{xx}(x,t) = \frac{\lambda}{h^2}[u(x+h,t+k) - 2u(x,t+k) + u(x-h,t+k)] +$$

$$+ \frac{(1-\lambda)}{h^2}[u(x+h,t) - 2u(x,t) + u(x-h,t)],$$

where λ is selected from [0, 1]. It is important to note that u_{xx} is approximated by a weighted average of the central difference derivatives in time t и $t+k$. At $\lambda = 0,5$ it turns out to be the usual average of these two central derivatives, and when $\lambda = 0,75$ – one of the difference derivatives is taken with a weight of 0.75, and the second – with a weight of 0.25. When $\lambda = 0$ you get the usual explicit scheme.

After replacing the partial derivatives u_t and u_{xx} in Eq. (15) the differential problem is obtained.

The differential equation is:

$$\frac{1}{k}(u_{i+1,j} - u_{i,j}) = \frac{\lambda}{h^2}(u_{i+1,j+1} - 2u_{i+1,j} + u_{i+1,j-1}) + \frac{(1-\lambda)}{h^2}(u_{i,j+1} - 2u_{i,j} + u_{i,j-1}),$$

$$\begin{cases} u_{i,1} = 0, \\ u_{i,n} = 0, \end{cases} \quad i = 1,2,...,m, \tag{16}$$

$$u_{1,j} = 1, \qquad j = 2,...,n-1.$$

Let us transfer all unknown values of the u from the upper temporary layer (with index i+1) to the left part of the Eq. (16), so we get

$$-\lambda r u_{i+1,j+1} + (1+2r\lambda)u_{i+1,j} - \lambda r u_{i+1,j-1} =$$

$$= r(1-\lambda)u_{i,j+1} + [1 - 2r(1-\lambda)]u_{i,j} + r(1-\lambda)u_{i,j-1}, \tag{17}$$

where $r = k/h^2$. Let us note that if i is fixed, and j varies from 2 to $n-1$ then Eq. (17) define the system of n-2 equations with n-2 unknown $u_{i+1,2}$, $u_{i+1,3}$, $u_{i+1,4}$,..., $u_{i+1,n-1}$, which are the solution of the problem in the inner grid knots on the temporary layer $t = (i+1)\Delta t\ \lambda$.

Figure 16.7 gives a visual representation of the structure of each equation of the system (the Eq. (17)). Let's proceed to the Eq. (17).

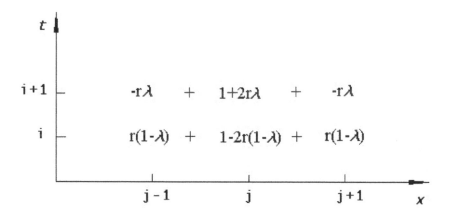

FIGURE 16.7   Template for implicit scheme.

Step 1. Choosing some value for $\lambda (0 \le \lambda \le 1)$. If $\lambda = 0$, then Eq. (17) transform to the explicit formulas from the previous section.

Step 2. Let, for example, $h = \Delta x = 0,2$ and $k = \Delta t = 0,08$ (in this case $r = \dfrac{k}{h^2} = 2$). In this case the grid contains 6 knots along the x-axis (4 internal knots). Taking the weight parameter (so that the scheme transform to a Crank-Nicholson scheme). In accordance with the computing template on Fig. 16.6 moving from left to right (j=2,3,4,5) on the first two layers (i=1), we obtain the following four equations:

$$-u_{21} + 3u_{22} - u_{23} = u_{11} - u_{12} + u_{13} = 1,$$
$$-u_{22} + 3u_{23} - u_{24} = u_{12} - u_{13} + u_{14} = 1,$$
$$-u_{23} + 3u_{24} - u_{25} = u_{14} - u_{15} + u_{16} = 1,$$
$$-u_{24} + 3u_{25} - u_{26} = u_{14} - u_{15} + u_{16} = 1.$$

They can be presented in a matrix form:

$$\begin{bmatrix} 3 & -1 & 0 & 0 \\ -1 & 3 & -1 & 0 \\ 0 & -1 & 3 & -1 \\ 0 & 0 & -1 & 3 \end{bmatrix} \begin{bmatrix} u_{22} \\ u_{23} \\ u_{24} \\ u_{25} \end{bmatrix} = \begin{bmatrix} 1 \\ 1 \\ 1 \\ 1 \end{bmatrix}. \tag{18}$$

The matrix of this system is called the three-diagonal matrix. For solving three-diagonal system,

$$\begin{bmatrix} b_1 & c_1 & 0 & 0 & \cdots & 0 \\ a_1 & b_2 & c_2 & 0 & \cdots & 0 \\ 0 & a_2 & b_3 & c_3 & \cdots & 0 \\ \vdots & \vdots & & & & \vdots \\ 0 & 0 & 0 & \cdots & a_{n-1} & b_n \end{bmatrix} \begin{bmatrix} x_1 \\ x_2 \\ x_3 \\ \vdots \\ x_n \end{bmatrix} = \begin{bmatrix} d_1 \\ d_2 \\ d_3 \\ \vdots \\ d_n \end{bmatrix},$$

let us convert it into an equivalent system

$$\begin{bmatrix} 1 & c^*_1 & 0 & 0 & \cdots & 0 \\ 0 & 1 & c^*_2 & 0 & \cdots & 0 \\ 0 & 0 & 1 & c^*_3 & \cdots & 0 \\ \vdots & \vdots & & & & \vdots \\ 0 & 0 & 0 & \cdots & 0 & 1 \end{bmatrix} \begin{bmatrix} x_1 \\ x_2 \\ x_3 \\ \vdots \\ x_n \end{bmatrix} = \begin{bmatrix} d^*_1 \\ d^*_2 \\ d^*_3 \\ \vdots \\ d^*_n \end{bmatrix},$$

where $\quad c^*_1 = \dfrac{c_1}{b_1}, \ c^*_{j+1} = \dfrac{c_{j+1}}{b_{j+1} - a_j c^*_j}, \ j = 1,2,...,n-2, \quad$ and

$$d^*_1 = \frac{d_1}{b_1}, d^*_{j+1} = \frac{d_{j+1} - a_j d^*_j}{b_{j+1} - a_j c^*_j}, j = 1,2,...,n-1 .$$

The new system is fully equivalent to the original one and its matrix is designed so that the system can be solved easily. Solving the equations sequentially from top to down, we can get

$$x_n = d^*_n, \ x_j = d^*_j - c^*_j x_{j+1}, j = n-1, n-2,...,2,1.$$

In the implicit scheme the volume of calculations at each step more than in the explicit one, but a good precision can be obtained even with a much larger step.

16.3.5 FINITE-DIFFERENCE NUMERICAL SHOCK-CAPTURING METHODS FOR CALCULATION OF TEMPERATURE FIELDS IN TWO-PHASE MEDIA

Any regular numerical scheme approximating Eqs. (3) or (4) can be applied for the calculation of continuous areas but schemes derived on the basis of the Eq. (1) are conservative and therefore more accurate. The implementation of conditions (the Eq. (5)) requires either the application of grid tied to the breakdown of the so-called floating grid or special calculation formulas for the cells near the surface of the break. Because of the complexities of software implementation in the two-dimensional and three-dimensional case this approach for the calculation of phase transitions is not usually used.

The second analytical supposes that breaks are not specifically identified and generalized solutions of the Eq. (1) are considered. Numerical methods based on this principle are called shock-capturing methods [2, 7]. The integral equation (the Eq. (1)) is always used for construction of difference schemes in these methods. Shock-capturing methods are divided into methods with regularization and without it. Regularization means that the calculation of discontinuous solutions is replaced with smooth solutions where breaks are replaced with narrow transition zones and a small parameter specifying the width of this zone is introduced. When this parameter converges to zero the width of a transition zone also converges to zero and so a discontinuous solution is got.

Temperature of the complete transition to the solid phase u_z different from zero is taken as a small parameter. Enthalpy on the segment $u_z < u < 0$ is a linear function of temperature. Such an approach allows to avoid possible problems with calculation stability related to the existence of solutions of integral equations.

16.3.6 GENERAL REQUIREMENTS FOR FINITE-DIFFERENCE NUMERICAL SCHEMES

After decomposition of the values in calculation cells in a Taylor series with respect to some calculation grid point and substitution of these values to the initial equation the approximation of this equation must be hold. The order of the remainder of such a substitution is the order of the approximation scheme. Calculation errors should not grow over time and this is checked experimentally or by using the spectral characteristic. As a rule, during the grinding of grid steps numerical solution gives the exact solution of the original equations in the limit. For linear equations it is proved in Ref. [2, 7] that convergence is achieved if the stability and approximation are hold. In nonlinear problems, which include phase transitions tasks, convergence is usually verified experimentally by grinding the grid and comparing the results with the known solutions. In the study of problems containing breaks [7, 10] (e.g., phase transition fronts and jumps of thermal conductivity coefficient), the conservative schemes should be used, that is, it is recommended to use the integrated conservation laws, in this case, as in the Eq. (1). At calculations of tasks with breaks and large gradients the property of monotony of the numerical scheme is efficient. This property means that the scheme does not create highs and lows that do not exist in the exact solution. In a nonmonotonic scheme with breaks and large gradients an effect of "saw" appears, that means formation of many highs and lows leading to a significant reduction of the calculation accuracy or even loss of stability due to overflow.

The factors described in this section are taken into account in the numerical scheme used in the software complex.

16.3.7 ANALYSIS OF METHODS AND ALGORITHMS, NUMERICAL SCHEME

Let us consider a conservative numerical scheme [7,12]. The calculation cell has the form of a parallelepiped. Mesh nodes corresponding to the vertices of the design volume are marked with integers. The values of temperature, thermal conductivity coefficients, heat capacity belongs to the midpoints of the cells, and these points are designated with half-integer indexes. The coordinates x_i, y_i, z_i, grid steps $h_{x,\,i+1/2} = x_{i+1} - x_i$, $h_{y,\,j+1/2} = y_{j+1} - y_i$, $h_{z,\,k+1/2} = z_{k+1} - z_k$ are specified in these nodes. After application of equation (1) to a design cell the equation is got:

$$h_{x,i+1/2}h_{y,j+1/2}h_{z,k+1/2}(H^{n+1}_{i+1/2,j+1/2,k+1/2} - H^{n}_{i+1/2,j+1/2,k+1/2}) =$$

$$= \{[h_{y,j+1/2}h_{z,k+1/2}(q_{i+1,j+1/2,k+1/2} - q_{i,j+1/2,k+1/2}) +$$

$$+ h_{y,j+1/2}h_{z,k+1/2}(q_{i+1/2,j+1,k+1/2} - q_{i+1/2,j,k+1/2}) +$$

$$+ h_{x,i+1/2}h_{y,j+1/2}(q_{i+1/2,j+1/2,k+1} - q_{i+1/2,j+1/2,k})] + h_{x,i+1/2}h_{y,j+1/2}h_{z,k+1/2}F_{i+1/2,j+1/2,k+1/2}\}\tau \qquad (19)$$

After division by $h_{xi}h_{yi}h_{zi}\tau$ we get a pattern for numerical schemes:

$$(H^{n+1}_{i+1/2,j+1/2,k+1/2} - H^{n}_{i+1/2,j+1/2,k+1/2}) / \Delta t =$$

$$= [(q_{i+1,j+1/2,k+1/2} - q_{i,j+1/2,k+1/2}) / h_{x+1/2,i} +$$

$$+ (q_{i+1/2,j+1,k+1/2} - q_{i+1/2,j,k+1/2}) / h_{y,j+1/2} + \qquad (20)$$

$$+ (q_{i+1/2,j+1/2,k+1} - q_{i+1/2,j+1/2,k}) / h_{z,k+1/2}] + F_{i+1/2,j+1/2,k+1/2}.$$

This is a difference scheme for a nonuniform grid, implemented in the software. The calculation is conducted in two stages: first flows are calculated, and then the enthalpy is calculated. It makes it convenient to formulate the second and third boundary conditions. Various conservative numerical methods differ from each other by different ways of the flow rate q calculating. Flows are calculated by approximation of Fourier's law:

$$q_{i,j+1/2,k+1/2} = \lambda_{i,j+1/2,k+1/2}(u_{i+1/2,j+1/2,k+1/2} - u_{i-1/2,j+1/2,k+1/2}) / h_i, \quad h_i = (h_{i+1/2} + h_{i-1/2})/2. \quad (21)$$

Here the flow approximation is given only in the x direction as an example, the remaining threads are approximated by the same way.

If are used than it leads to an explicit scheme. Such scheme is used in the applicable software. And an implicit scheme appears if use the temperature values from the upper layer. An explicit scheme requires the stability condition, which can be derived experimentally or by the spectral characteristic. In the case of the heat equation with unit coefficients and the same steps on all spatial variable for the explicit scheme the stability condition becomes $\tau \leq 0.5(h_x^2 + h_y^2 + h_z^2)/3$. Implicit schemes are more resistant, but purely implicit two-layer scheme is absolutely stable. Implicit schemes require special algorithms for solving implicit equations [2, 10]. During the computational realization systems of implicit equations are solved by iterative algorithms, which often are reduced to the solution of the heat transfer equation using explicit schemes. With uniform grids applying it is usually not justified because of the large consumption of machine time comparable with calculations using explicit schemes with a small step. Calculations by the explicit schemes at the expense of small steps are more accurate. Introduction of implicit schemes complicates the modification of programs and the incorporation of the various additional blocks. But if there are areas with substantially different values of thermal conductivity and heat capacity or in case of nonuniform grids in the presence of regions with significantly different spatial steps the application of implicit schemes is appropriate.

In the class of implicit schemes there is a scheme of the second order accuracy in time, where the temperature values are taken as the half-sum of the values from the upper and lower layers. Such a scheme is not monotonous and wasn't used during the software package creation. For the same reason an explicit three-layer Dufort-Frankel scheme (rhombus) was rejected. In this case the time difference is calculated with values through two layers and the temperature flows in the central point are calculated as a half-sum of the values on the lower and the upper layer [2, 7]. This scheme is absolutely stable but requires a small time step for accuracy achievement.

Similar schemes [7, 11] exist for the Eq. (3). However, calculations of phase transitions using Eq. (3) always need regularization and a small time step. Because of a quasilinear character of Eq. (3) systems of linear equations appear when using implicit schemes. This allows to build an implicit scheme called a double-sweep method of variable directions. Consumption of machine time for realization of one time step here is of the same orders that in the explicit schemes. However, the inclusion of the additional calculated objects, which should be provided with the boundary conditions in this scheme, makes the algorithm much more complicated. For this reason explicit schemes are used in the software.

Schemes also differ in the way of calculating of the thermal conductivity coefficient. Different ways are used. One of them is to give these factors immediately on the borders. It is known that this method may lead to fluctuations at breaks. So usually flattening is used, for example, coefficients in the middle of the nodes are specified and missing values on borders are found by linear interpolation.

The software considers a method, which is optimal for calculation of thermal conductivity coefficients [7]:

$$q_{i,j+1/2,k+1/2} = 2(u_{i+1/2,j+1/2,k+1/2} - u_{i-1/2,j+1/2,k+1/2}) / \left(\frac{h_{x,i+1/2,j+1/2,k+1/2}}{\lambda_{i+1/2,j+1/2,k+1/2}} + \frac{h_{x,i-1/2,j+1/2,k+1/2}}{\lambda_{i-1/2,j+1/2,k+1/2}} \right). \quad (22)$$

This approximation is optimal in the sense that the conditions on the break contact between two computational cells are performed precisely, however, the use of this method is not overemphasized. In a circle of experts on heat tasks schemes constructed with the use of such an approximation flows sometimes are called the "method of control volume."

Differential methods in which the spatial steps depend on the provisions of the computation node in space are called irregular. In the simplest case discussed above:

$$h_x = h_x(x), \quad h_y = h_y(y), \quad h_z = h_z(z) \quad (23)$$

A more sophisticated variant:

$$h_x = h_x(x, y, z), \quad h_y = h_y(x, y, z), \quad h_z = h_z(x, y, z) \quad (24)$$

Irregular grids allow to bind the grid to calculation objects and realize the computation in certain areas with small steps. However, this is not a universal method [7]: uniform meshes usually have a higher order of accuracy of approximation in space, that is why at a sufficiently fine grid snap to objects is becoming irrelevant [2]. The first option (the Eq. (23)) is most effective in one-dimensional calculations, in two-dimensional and three-dimensional cases this approach leads to the situation when step becomes fine in areas where it is not required. The second option (the Eq. (24)) has been developed in recent years [9–12], it requires significantly more complex algorithms for its implementation. One of the methods is that the calculation step to the next cell consistently decreases two times.

This can be done automatically with increasing of temperature gradient. Also a combined approach can be applied: uniform grid is spliced with a fine mesh in a certain areas.

In the case of uniform grids explicit scheme acquires a more simple form. For a homogeneous environment:

$$\frac{H_{i,j,k}^{n+1} - H_{i,j,k}^{n}}{\lambda \Delta t} = \frac{u_{i+1,j,k}^{n} + u_{i-1,j,k}^{n} - 2u_{i,j,k}^{n}}{\Delta x^2} + \frac{u_{i,j+1,k}^{n} + u_{i,j-1,k}^{n} - 2u_{i,j,k}^{n}}{\Delta y^2} +$$

$$+ \frac{u_{i,j,k+1}^{n} + u_{i,j,k-1}^{n} - 2u_{i,j,k}^{n}}{\Delta x^2} + F_{i,j,k}.$$

(25)

For a heterogeneous medium:

$$\frac{H_{i,j,k}^{n+1} - H_{i,j,k}^{n}}{\Delta t} = \frac{\lambda_{i-1/2,j,k} \dfrac{u_{i-1,j,k}^{n} - u_{i,j,k}^{n}}{\Delta x} + \lambda_{i+1/2,j,k} \dfrac{u_{i+1,j,k}^{n} - u_{i,j,k}^{n}}{\Delta x}}{\Delta x} +$$

$$+ \frac{\lambda_{i,j-1/2,k} \dfrac{u_{i,j-1,k}^{n} - u_{i,j,k}^{n}}{\Delta y} + \lambda_{i,j+1/2,k} \dfrac{u_{i,j+1,k}^{n} - u_{i,j,k}^{n}}{\Delta y}}{\Delta y} +$$

$$+ \frac{\lambda_{i,j,k-1/2} \dfrac{u_{i,j,k-1}^{n} - u_{i,j,k}^{n}}{\Delta z} + \lambda_{i,j,k+1/2} \dfrac{u_{i,j,k+1}^{n} - u_{i,j,k}^{n}}{\Delta z}}{\Delta z} + F_{i,j,k}.$$

(26)

Here, for simplicity of notation, the numerical scheme temperature attributed to integer nodes and coefficients of thermal conductivity – to half-integer ones.

16.3.8 APPROXIMATION OF BOUNDARY CONDITIONS

The simplest method of approximation of boundary conditions of the first type for a surface perpendicular to the x-axis for the numerical scheme shown above is the following:

$$u_{i+1/2, j+1/2, k+1/2} = \mu.$$

(27)

The boundary condition is delivered in a half-integer node with index $i+1/2$. It is immaterial if the grid is shallow or if this condition is exposed on the bottom of the design volume. If necessary, this condition can be put in the node with an integer index i using another method:

$$q_{i,j+1/2,k+1/2} = 2(\mu - u_{i-1/2,j+1/2,k+1/2}) \frac{\lambda_{i-1/2,j+1/2,k+1/2}}{h_{i-1/2}},$$

$$q_{i,j+1/2,k+1/2} = -2(\mu - u_{i+1/2,j+1/2,k+1/2}) \frac{\lambda_{i+1/2,j+1/2,k+1/2}}{h_{i+1/2}}. \tag{28}$$

In the first case, the calculated volume is on the left of the surface, in the second case – on the right.

The boundary condition of the second type in the node with index i looks like the following:

$$q_{i,j+1/2,k+1/2} = V, \; q_{i,j+1/2,k+1/2} = -V. \tag{29}$$

Homogeneous boundary condition of the second type in the nodes with the integer index i can be shown using the temperature by adding additional boundary nodes:

$$u_{i+1/2,j+1/2.k+1/2} = u_{i-1/2,j+1/2.k+1/2}. \tag{30}$$

This condition is also the condition of the symmetry: it is used for calculations using symmetry properties of the problem statement.

The boundary condition of the third kind in the node with index i can be presented as:

$$q_{i,j+1/2,k+1/2} = 2(\mu - u_{i-1/2,j+1/2,k+1/2}) \frac{2}{h_{i-1/2}/\lambda_{i-1/2,j+1/2,k+1/2} + 2(1/\alpha_{sur} + R_{sur})},$$

$$q_{i,j+1/2,k+1/2} = -2(\mu - u_{i+1/2,j+1/2,k+1/2}) \frac{2}{h_{i+1/2}/\lambda_{i+1/2,j+1/2,k+1/2} + 2(1/\alpha_{sur} + R_{sur})}. \tag{31}$$

Here R_{sur} is a coefficient of thermal resistance of the surface, α_{sur} is a coefficient of convective heat transfer.

16.4 SIMULATION OF TEMPERATURE FIELDS IN A MATERIAL WITH EMBEDDED NANOTUBES

In modern nanotechnologies many objects of quantum physics are used such as pillboxes, quantum wires, super lattices, etc. One interesting feature of the structural elements are nanotubes. Nanotubes are long molecules consisting of a large number of atoms placed on the cylindrical shaped spatial surfaces. Currently various nanotubes are synthesized, including those alloyed with metal.

Two-phase nanomaterial is taken as a design volume. Characteristic sizes and temperature originally are nondimensionalized, it allows us to reduce the problem to the universal type. Nondimensionalizing is carried out so that a critical temperature (temperature of the phase transition) equals to 0.

The first stage of calculations implies observation of temperature distribution changing in nanomaterial in the presence of the external surface heat source. Spatial calculation steps are equal on all axes. It is considered that within this area at the upper limit of the design volume the first boundary condition – a constant temperature – will take place. Let us consider the temperature changes in the structure of nanomaterial over time. Figs. 16.8 and 16.9 show the graphs of temperature changes in the observation areas. The first, the second and the third points are located on the axis of symmetry of the, on the corner of the design volume and in the field without the influence of external heat, respectively. Point 4, 5 and 6 are selected in the same areas but in contrast to the points 1, 2 and 3 they are located deeper in the volume of the nanomaterial.

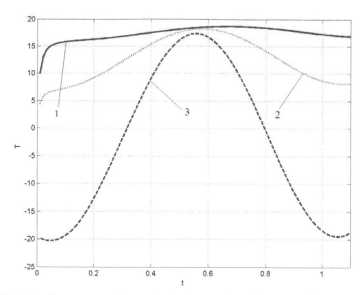

FIGURE 16.8 Temperature changes in the observation points 1, 2, 3 for a certain period of time.

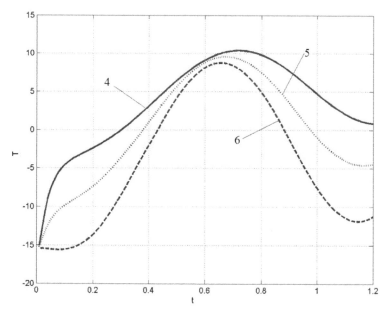

FIGURE 16.9 Temperature changes in the observation points 4, 5, 6 for a certain period of time.

Figure 16.10 reflects the cyclical changes of temperature distribution in the structure of nanomaterial if the surface temperature changes in time.

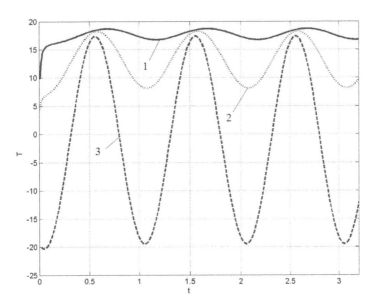

FIGURE 16.10   Cyclical temperature changes in observation points 1, 2, 3 over time.

The next stage considers carbon nanotubes with the refrigerant core, which are embedded in the nanostructure and positioned as heat outflows. In the numerical realization four nanotubes were considered to be embedded in the material on the corners of the area exposed to external heating. A nanotube was programmed as an infinitely thin source (outflow) of heat with constant power. Let us consider the temperature profiles in the observation areas. The observation points were taken similarly to the first stage of the study. Figure 16.11 presents the results of numerical simulation for the temperature distribution in points 1, 2 and 3 which spatial distribution was mentioned above. It is obvious that the heat outflows influence the temperature of nanostructure: if earlier the design volume warmed up to the phase transition temperature, it can be seen that now the temperature on the axis of symmetry does not exceed the critical value.

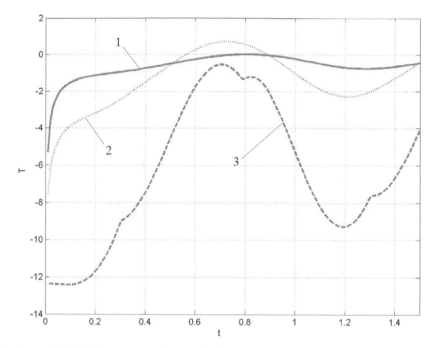

FIGURE 16.11   Temperature changes in the observation points 1, 2, 3 over time.

Finally, let us compare the values of temperature in the design volume with and without usage of nanotubes. Fig. 16.12 shows temperature profiles of points located on the axis of symmetry (point 1) and at the corner of the design volume (point 2). Black color highlights temperature curves in the case when nanotubes are not applied. Red color shows temperature profiles of points in nanostructure with embedded nanotubes positioned as heat outflows. Here temperatures have a sinusoidal shape with a maximum in the neighborhood of the critical temperature. Only for the point at the corner of the design volume a slight temperature increase can be noticed after the period of external heating, which can be avoided using a larger number of heat outflows or their different geometrical superposition.

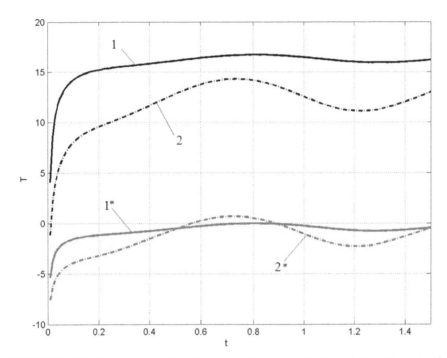

FIGURE 16.12   Temperature values in observation points: 1, 2 – points on the axis of symmetry and on the corner of the design volume (without heat outflows); 1*, 2* – with heat outflows.

Also a study considering the horizontal implementation of nanotubes was hold. The possibility of periodic functioning of nanotubes was taken into account (the period of work is when the refrigerant moves in cores of nanotubes providing thermo stabilizing influence on the distribution of heat in the nanostructure). Fig. 16.13 presents a comparison of temperature dynamics in observation points 1 and 2 during the periodic cycle of work of nanotubes.

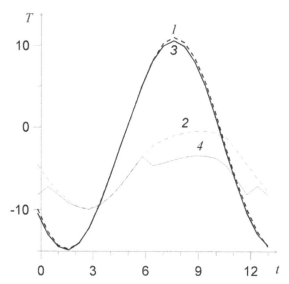

FIGURE 16.13 MCalculation considering a horizontal nanotube with the shift cycle of functioning: *1, 2* – the temperature in the first and second point of observation without nanotubes, *3, 4* – temperature at the first point and the second point during the application of nanotubes.

16.5 CONCLUSIONS

On the basis of the theory of numerical methods and mathematical modeling the problem of the calculation and forecast of the distribution of the temperature field in a two-phase nanocomposite environment is solved. The mathematical statement of the problem is formulated as the integral equation of thermal balance with a heat flux taken into account, which changes according to Fourier's law. Jumps of enthalpy and heat conductivity coefficient are considered. Various numerical schemes and methods are examined and the best one is selected – the method of control volume. Calculation of the dynamics of the temperature field in the nanostructure is hold using the software.

The results of calculations and forecast of the temperature fields distribution in nanostructures with a variation of the nanotubes' implementation parameters are presented. A positive effect on the temperature field of nanomaterial and the ability to control the heat dynamics to prevent critical temperature values assuming the phase transition are demonstrated.

KEYWORDS

- **control**
- **mathematical simulation**
- **nanocomposite**
- **nanotubes**
- **numerical methods**
- **temperature field**

REFERENCES

1. Belov, V. V., Dobrohotov, S. U., Tudorovskia, T. J., Asymptotic solutions of nonrelativistic equations of quantum mechanics in curved nanotubes. Theor. Math. Phys., 2004, V. 141, №2, pp. 267–303.
2. Samarskiy, A. A., Gulin, A. V., Numerical methods. Moscow: Nauka, 1989, 432 p.
3. Klimov, D. M., Vasiliev, A. A., Luchinin, V. V., Malyacev, P. P., Prospects of development of Microsystems engineering in the XXI century Nano – and microsystem technology, 1999. № 1. pp. 3–6.
4. Kosobudski, I. D., Ushakov, N. M., Urkov, G. U., Introduction to the chemistry and physics of nano-sized objects-Saratov: SGTU, 2007. 182 p.
5. Krylov, D. A., Sidnyaev, N. I., Control volume method in problems of temperature forecasting in three-dimensional areas Abstracts of papers. University research conference "Student spring – 2011." Moscow, BMSTU, 2011. Vol. XI, part 1, pp. 125–127.
6. Bakiev, T. A., Heat exchangers development on basis of thermosiphons for oil production and refining PhD dissertation (Technics). Ufa State Petroleum Technological University, Ufa, 2000, 228 p.
7. Patankar, S. V., Computation of conduction and Duct Flow Heat Transfer. Moscow: MEI, 2003, 312 p.
8. Samarskiy, A. A., Vabischevich, P. N., Computational heat transfer. Moscow, Book house "Librocom," 2009, 784 p.
9. Krylov, D. A., Melnikova, Y. S., Mathematical modeling of temperature fields distribution in permafrost. Student scientific journal. Collection of articles of the fourth scientific and technical exhibition "Polytechnic." Moscow, BMSTU, 2009, pp. 94–97.
10. Sidnyaev, N. I., Hrapov, P. V., Melnikova, Y. S., Principles of mathematical modeling of temperature fields in multiphase media. Collection of reports of IV all-Russian

youth scientific-innovative school "Mathematics and mathematical modeling" 19–22
April 2010, G. Sarov, 2010, pp. 85 93.

11. Krylov, D. A., Melnikova, Y. S., Mathematical modeling of temperature fields distribution in permafrost. Student scientific journal. Collection of articles of the fourth
scientific and technical exhibition "Polytechnic." Moscow, BMSTU, 2009, pp. 94–97.

12. Sidnyaev, N. I., Fedotov, A. A., Melnikova, Y. S., Controlling of temperature fields
distribution in permafrost. Academia. Architecture and building. Moscow, NIISF
RAASN, #3, 2010, pp. 372–374.

13. Sidnyaev, N. I., Theory of phase transitions and statistical phenomena mechanics of
nanostructured materials Vestnik MGTU. Special issue "Nanoingeneriya." Instru-
ment-making. 2010, pp. 9–22.

14. Dexter J. Carbon Nanotubes Enable Pumpless Liquid Cooling System for Comput-
ers IEEE Spectrum's nanotechnology blog. URL: http: spectrum.ieee.org/nanoclast/
semiconductors/nanotechnology/carbon-nanotubes-enable-pumpless-liquid-cooling-
system-for-computers.

CHAPTER 17

INFLUENCE OF NANOTECHNOLOGY ON COILED SPRINGS OPERATIONAL CHARACTERISTICS

O. I. SHAVRIN

CONTENTS

ABSTRACT

This chapter describes the technology of nanoscale substructure forming in coiled springs made by hot forming. Research results of the technology effect on coiled springs operational characteristics that is, durability and spring camber resistance under cyclic load are given.

17.1   INTRODUCTION

For structural materials and products that are made of them, nanotechnology has to be understood as the formation processes of any structure parameters that get nanoscale dimensions corresponding to the established requirement – less than 100 nm, at least in one of the measuring directions. Strengthening effect of nanoscale structure parameters is dislocation impediment ensuring increasing strain resistance under stress. These structure parameters can be grain boundaries, dispersive carbide, nitrides, carbonitrides and other precipitates, subgrain boundaries resulting from reconfiguration of dislocations that are formed during plastic strain applied either to metal or part production. The only reason to study the processes of metal and part production lies in the fact that to get a metal with any special properties such as high metal strength and hardness is one problem, the other one, is to make parts of this metal with acceptable efficiency.

Therefore, from the point of view of nanotechnology implementation in the production of engineering products, development of metal nanoscale structure element technologies in the finished part is the most urgent matter.

Considering the possible link between the concepts "real metal," "a product made of it" and "nano" one can offer a unique algorithm of problem examination.

17.1.1   AIM TO BE ACHIEVED

One of the aims can be improvement of operational characteristics. The variety of options for the concept "operational characteristics" dictates the appropriate means to achieve goals. Speaking just about engineering parts, one can limit a set of operational characteristics at least by structural strength. Structural strength is the type of loading (static, impact, cyclic), temperature operational conditions, the ratio of operational stress level and metal strength characteristics, the surface condition of a part, stress concentrators, etc.

17.1.2 STRUCTURAL FACTORS DETERMINING THE OPERATIONAL CHARACTERISTIC LEVEL

The role of structural factors is determined by the type of the operational loading that provides different structural factors of the required product efficiency level.

The aforesaid suggests that the developing scheme to realize or obtain "nano" for real metals and structures should be aimed at getting some specific material structure property determining the efficiency of a part under the operational conditions. Let's consider whether it is possible to link the concept "nano," that is, the possibility of getting changes in the structure of part material at the nano level that undergoes cyclic loading or fatigue.

One can distinguish two independent structural factors — the size of structural elements and carbide particles formed by the technology for such class of parts from a large number of factors that determine the part efficiency under cyclic loading.

Strengthening role of carbide dispersion is known. It is also known that getting carbides with the size of 20–100 nm is possible by conventional quenching of structural steels [1].

For parts under fatigue the most important thing is the size of substructure elements to increase the level of failure stress. This hypothesis is based on the established facts.

1. The process of fatigue cracks origin results from the formation of critical dislocation clusters in front of structural barriers.
2. Operational stresses triggering dislocations movement constitute a portion of the elastic limit.
3. Polygonal subboundaries and cells inside grains become effective barriers for dislocation movement under these stresses. The dispersity of substructure elements initiates the effect of dislocation

movement steadiness and difficulty of critical density forming increasing the number of loading cycles required [1].

Studying the influence of subgrain sizes on strength it was found out that the Hall-Petch empirical dependence for grain sizes in pure metals without substructure is

$$\sigma_T = \sigma_o + k_y D^{-1/2}$$

where σ_o is stress that should be applied to overcome the friction of the lattice (of Peierls-Nabarro forces); k_y is quantity that characterizes the degree of dislocation locking and braking, applicable for metals with the developed substructure even at the change of subgrain sizes from 0.05 micron (50 nm) to 70 microns, when subgrain sizes are substituted in the formula instead of the grain sizes. It was also shown in the works of M.L. Bernstein and other scientists when studying the substructure formation during the process of high-temperature thermomechanical treatment (HTMT) (steel strains at temperature of ≈ 1000°C followed by quenching) for sample alloys preserving austenitic state with the characteristics of the fine structure formed under the influence of high-temperature deformation after cooling to room temperature. Results of these studies showed that the polygonal subboundaries can contribute to hardening that exceeds hardening caused by grain boundaries. Dislocation density has a significant effect on hardening increased manifold by thermomechanical treatment.

This is also shown in Ref. [2], where it is proved that in addition to the simple dependence on grain, there is an additional effect due to subgrains surrounded by small-angle subboundaries being the obstacles for dislocations and contribution to the increase of the flow stress. Small-angle subboundaries increase friction during dislocation movement inside the grain without corresponding change its size.

In Refs. [1] and [3], it is shown that created in HTMT substructure provides more effective resistance to small plastic strains under loading, that is, as long as it does not fail as a result of significant strains under high stresses.

The dimension of the polygonal substructure in HTMT is determined by the parameters of all process phases, that is, temperature and heating method, the degree and strain mode, and method and scheme of cooling.

The type and dimension of the forming substructure are influenced by the strain duration and the pause between straining and cooling. The dimension of the original structure [4] is also of great importance. The formation of a polygonal nanoscale substructure is possible at certain HTMT process parameters. At the optimal heating temperature ensuring the homogeneity of high-temperature phase, the substructure dimension is influenced by interrelated strain amount and the duration of after strain interval before quenching.

Studies of possibility to form the nanoscale substructure in metals of machine components and its influence on operational characteristics were carried out on coiled springs made by hot forming. Such a spring is a suitable object for structuring process simulation under heat-induced strain: spring winding is done under high temperatures that allow quenching under winding heating, geometrical parameters provide for sufficient (10–20%) strain amount within the outer metal layers of the spring to achieve heat-induced strain effect. Taking into account that strain amounts are less than generally accepted ones for HTMT, this process is called small-strain thermomechanical treatment (SSTMT).

SSTMT implementation technology for coiled springs includes the following operations: rod heating, spring winding and quenching with mandatory after winding (after strain) interval. The effect of this technology on spring material structure was studied for a number of designs of bogie springs of freight railroad cars with various $C = \dfrac{D}{d}$ factor, where D is the outer spring diameter, d is rod diameter.

Material hardness of the produced springs corresponded to the standard requirements for springs of railroad rolling stock of 42 … 46 HRC.

Steel structure of such hardness is troostite.

The fine structure of spring material was defined under electron microscope investigation. Researches were carried out on transmission electron microscope ЭM-125 K under accelerating voltage of 100 kV and 40,000 … 300,000 times amplification.

During electron microscope investigation Fig. 17.1 it was founded that the fine material structure of springs produced by SSTMT method was uniform with characteristic increased density of dislocations, subboundaries, disperse equally distributed carbide particles. Average ferrite matrix substructure elements constituted 20 … 40 nm at the dispersion of up to 100 nm.

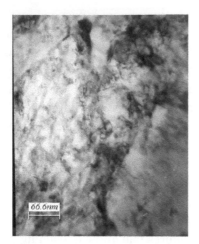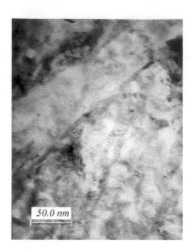

FIGURE 17.1 Fine material structure of coiled springs (Steel 60C2A (C – 0.57–0.65%, Si – 1.5–2.0%), reheat temperature = induction heating –1020°C, strain 20%, tempering 460°C).

Electron diffraction patterns prove the small-angle disorientation of substructure elements.

Steel fine structure after quenching with zero strain value is characterized by the absence of substructure, decoration of crystal boundaries with carbides that are coarser than the carbides inside the crystals.

Spring steel produced by SSTMT method is characterized by higher carbide degree of dispersion, their sizes are within the limits 7 … 10 nm.

These results show that the substructure formed under hot working during spring winding and quenching specified cooling, inherited by martensite, is retained in steel structure after tempering.

It is possible to speak about the influence of material substructure of a spring on operational characteristics by spring camber research data, relaxation resistance under cyclic loads, fatigue life test and the failure stress values.

These characteristics were studied for coiled springs of spring suspension for various types of bogies used in freight cars of Russian Railways.

The springs were manufactured by means of controlled nanosubstructure formation method in material due to SSTMT.

Some results of safety regulation tests adopted by Russian Railways JSC are given in Table 1. Springs were made of steel 60C2ХФА (C – 0.56–0.64%; Si – 1.4–1.8%, Mn – 0.4–0.7%, Cr – 0.9–1.2%, V – 0.1–0.2%).

The results given in Table 17.1 indicate the high quality of the tested springs and stability of their geometrical and load bearing characteristics. Results in terms of index 2 – spring coil clearance stability and zero spring camber under triple compression are crucial as they speak for high microflow strength of nanostructured steel. Uniformity of spring coil clearance is of primary importance for spring operation as it eliminates the possibility of interlocking of spring coils and early fracture of springs.

TABLE 17.1   Results of safety regulation tests.

No.	Index	Unit measure	Index value			
			Spring in Fig. 7		Spring in Fig. 6	
			allowable	control	allowable	control
1.	Limiting deviation per spring free height	mm	+7.00	+4.40	+7	+4.53
			−2.00	+4.75	−2.00	+3.97
				+3.00		+4.03
2.	Limiting deviation per difference between maximum and minimum spring coil clearance	mm	7.75	0.30	5.25	0.8
				1.00		0.5
				0.30		1.2
3.	Limiting compression deviation under static load	mm	+8.00	+4.00	+8.00	+3.00
			−5.50	+1.00	−5.50	+2.00
				+2.00		+1.00
4	Residual strain after triple compression till contact of coils	mm	2.00	0.1	2.00	0.0
				0.2		0.2
				0.2		0.1

Data given in Figs. 17.2 and 17.3 also indicate the same. The data in Fig. 17.2 show that when heating springs of similar design but made by different technology have different loading diagrams – springs made by conventional technology suffer from microflow under ultimate loads (till contact of coils), while springs with material nanosubstructure show constant linear relationship of load – compression.

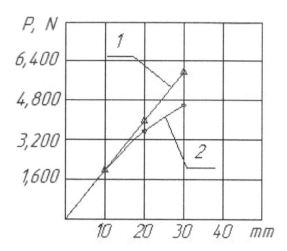

FIGURE 17.2   Compression diagram of springs made of steel 60C2 (C – 0.57–0.65%, Si – 1.5–2.0%). [1(Δ) – SSTMT spring, $T_{heating}$ = 1020°C, $T_{tempering}$ = 460°C, outer fibers strain amount 15%, 2(O) conventionally heat treated spring, $T_{quenching.}$ = 860°C, $T_{tempering}$ = 460°C. Spring parameters: number of coils – 7, spring coil clearance – 4.5 mm, rod diameter – 10 mm, rigidity index c = 5.]

Test results of spring relaxation resistance under cyclic loads also prove higher resistance to microflow of spring material with nanosubstructure (Fig. 17.3). Spring free height changing was measured during fatigue test. As it is evident from the diagrams the camber value of springs with nanoscale structure was 10 times less than the springs made by conventional technology.

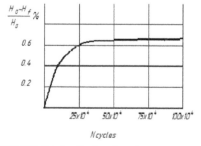
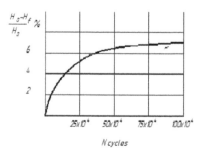

FIGURE 17.3   Spring free height changing H_0 (spring camber) under cyclic loading: a) SSTMT springs, $T_{heating}$ = 1020°C, $T_{tempering}$ = 460°C, outer fibers strain amount 15%, b) conventionally heat treated springs, $T_{quenching}$ = 860°C, $T_{tempering}$ = 460°C. Steel 60C2A (C – 0.57–0.65%, Si – 1.5–2.0%) testing stress 900 MPa.

Springs were made of steel 60C2A (C – 0.57–0.65%, Si – 1.5–2.0%). Testing loading was at pulsating stress. The picture caption states the outer fibers strain amount, which is the elongation of the outer fiber under spring winding. It depends on the value of factor $C = \dfrac{D}{d}$.

Certification tests were carried out for springs applied in various freight car bogie designs on account of the beginning of coiled springs with formed nanoscale substructure manufacturing for freight car spring suspension.

Test results were compared with the results obtained from spring testing made by conventional hot forming with quenching under winding heating (Figs. 17.4 and 17.5). Bogie coiled springs (Figs. 17.6 and 17.7) are made of steel 60C2ХФА (C – 0.56–0.64%; Si – 1.4–1.8%, Mn – 0.4–0.7%, Cr – 0.9–1.2%, V – 0.1–0.2%). Each cushioning element has two springs one inside the other – internal and external ones.

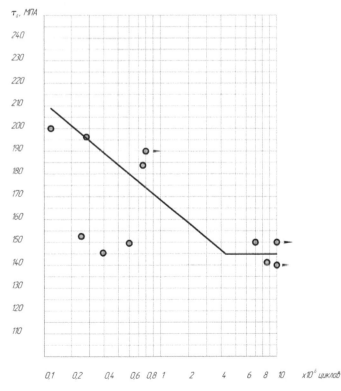

FIGURE 17.4   Spring fatigue test results (drawing Fig. 17.6, hot forming technology).

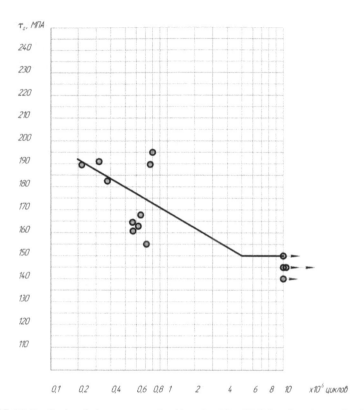

FIGURE 17.5   Spring fatigue test results (drawing Fig. 17.7, hot forming technology).

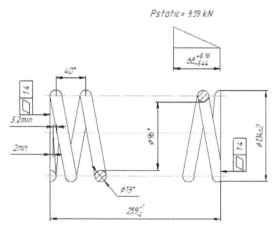

FIGURE 17.6   Spring drawing, steel 60С2ХФА (C – 0.56–0.64%; Si – 1.4–1.8%, Mn – 0.4–0.7%, Cr – 0.9–1.2%, V – 0.1–0.2%).

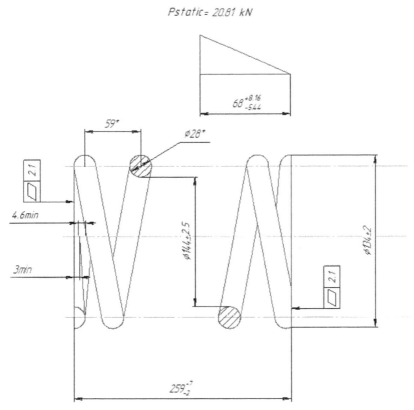

FIGURE 17.7   Spring drawing, steel 60C2XΦA (C – 0.56–0.64%; Si – 1.4–1.8%, Mn – 0.4–0.7%, Cr – 0.9–1.2%, V – 0.1–0.2%).

Test results of such springs made by the technology of nanoscale sub-structure forming inside the spring material are given in Table 17.2 and Fig. 17.8. Spring loading was carried out in correspondence with the drawing requirement in terms of asymmetrical repeated stress cycle with the static component assigned by the drawing. Spring static strain amount is given in the second column of the Table and constitutes 68 mm. Variable strain component is given in column 3. Operational strain variable for the spring in Fig. 17.4 is assigned ±23 mm, while for the spring in Fig. 17.5 it is ± 21 mm.

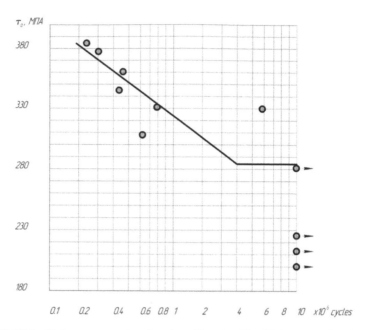

FIGURE 17.8 Fatigue test results of springs (drawing Fig. 17.6, nanoscale substructure forming technology (SSTMT).

TABLE 17.2 Test results

Spring No.	Strain static component, mm	Strain variable component, mm	Stress variable component, MPa	Number of cycles before rupture	Note
1	68	27	235.8	10×10^6	Without rupture
2	68	27	235.8	10×10^6	Without rupture
3	68	27	225.0	0.8×10^6	
4	68	26	207.6	10×10^6	Without rupture
5	68	26	214.1	10×10^6	Without rupture
6	68	27	202.1	1.67×10^6	
7	68	27	215.6	10×10^6	Without rupture
8	68	27	223.3	10×10^6	Without rupture
9	68	27	215.6	10×10^6	Without rupture
10	68	27	208.8	10×10^6	Without rupture

TABLE 17.2 *(Continued)*

Spring No.	Strain static component, mm	Strain variable component, mm	Stress variable component, MPa	Number of cycles before rupture	Note
11	68	25	193.4	10×10^6	Without rupture
12	68	25	193.4	10×10^6	Without rupture
13	68	21	178.1	10×10^6	Without rupture
14	68	21	162.4	10×10^6	Without rupture

Attempts at pursuing high stability for the springs (Fig. 17.5) made by nanoscale substructure forming technology have failed due to insufficient test-bench power meant for testing springs made by conventional hot forming technology. For ordinary springs it is possible to take stress value equal 190 MPa (fatigue curve, Fig. 17.5) number of cycles to failure $\approx$ 230,000 for springs with nanoscale structure it is 10×10^6 cycles to failure to compare their strength levels.

For springs (Fig. 17.6) and springs with nanoscale structure building fatigue curve (Fig. 17.8) was successful. Comparison of the obtained curves shows the difference between failure stress values under the same testing conditions for springs made by conventional hot forming technology and springs made by nanoscale substructure forming technology.

From the given data it is possible to make a conclusion:

1. Application of SSTMT method for coiled springs manufacturing forms allows forming the nanoscale substructure in spring material.

2. Springs with nanoscale substructure have greater operational characteristics.

3. Operational characteristics level of spring with nanoscale substructure allows design of new generation coiled spring structures with reduced mass.

KEYWORDS

- coiled spring
- durability
- nanoscale substructure
- relaxation resistance
- strength

REFERENCES

1. Bernstein M. L., Dobatkin S. V., Kaputkina L. M., Prokoshkin S. D. Hot working diagrams, steel structure and properties. – Moscow: "Metallurgia," 1989. 528 p.
2. English A. T., Bakofen U. A. Influence of metal working on rupture. 1983. 176 p.
3. Shavrin O. I. Methods and equipment for thermomechanical treatment of machine components. – Moscow: "Mashinostroenie," Moscow: Metallurgia, 1976
4. Shavrin O. I. Forming of nanoscale substructure in the material of machine components. ISTU Vestnik No.1, 2011, pp. 4–7.
5. Shavrin O. I. High strength springs for railway rolling equipment. "Industrial transport XXI century" No. 3, 2012, pp. 16–18.

CHAPTER 18

DRAWING OF CONTINUOUS PROFILES OF NOT ROUND CROSS SECTION FOR PRODUCTION OF COMPOSITE LOW-TEMPERATURE SUPERCONDUCTORS

V. N. TROFIMOV and T. V. KUZNETSOVA

CONTENTS

ABSTRACT

The perspective direction of development of electrical equipment is production of composite electroconductors, for example, low-temperature superconductors in which the effect of superconductivity is caused by effect of a pinning – fixing of whirlwinds of a magnetic field on the defects of crystal structure being objects of a nanolevel. Carriers of such defects is the fibrous structure received at plastic deformation with big extents of deformation.

For receiving demanded fibrous structure the production technology of low-temperature superconductors includes repeated drawing of bars or a wire of not round section a method, their cutting, formation of composite preparation and preparation drawing to a given size.

With use of model of an ideal plastic material formulas for calculation of tension of drawing, definition of optimum geometry of the channel of the drawing tool and extent of plastic deformation are received when drawing continuous profiles of not round cross section.

18.1   INTRODUCTION

Lengthy profiles of not round cross-section section from color and ferrous metals and alloys are widely applied in different areas in the industry. For example, in mechanical engineering the calibrated steel rectangular profiles are used for manufacturing prismatic elements. In instrument making and the electrical engineer aluminum and copper electroconductors of a rectangular profile as allow to receive high degree of filling of section at winding of coils of transformers, electromagnets, solenoids, etc. are used

Perspective direction of development electrical engineers is application of composite electroconductors, in particular, low-temperature superconductors (LTSC) for superconducting magnetic systems [1].

Technological process of manufacture LTSC includes reception bars or a wire correct six-sided sections from Nb, copper, alloy NbTi and bronze, them cutting on pieces of the set length and formation of composite preparation (Fig. 18.1). At formation of composite preparation for increase of density of filling of peripheral zones of section of a composite use bars rectangular or round section. As a result of multiphasic plastic deformation till the set diameter the fibrous structure is formed. On Fig.18.2, the section of a fiber of one of kinds LTSC in diameter 5 microns is shown. The effect of superconductivity is provided for the account of formation of structure of grains intermetal Nb_3Sn, which is formed after deformation of composite preparation and the diffusive annealing. It is visible that grains Nb_3Sn have the size of 10–100 times of less diameter of a fiber, that is are nano objects.

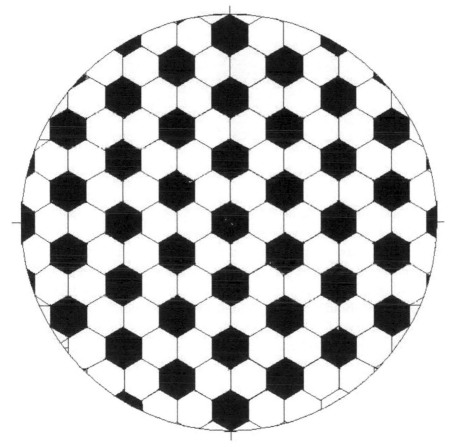

FIGURE 18.1   Section of composite preparation.

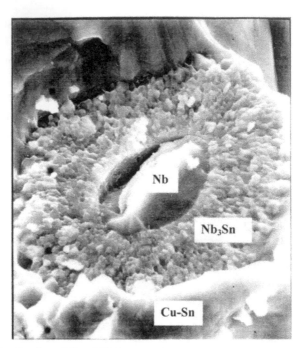

FIGURE 18.2   Section of fiber LTSC.

The technology of manufacture of lengthy profiles of not round section should exclude breakage preparations and provide the minimum density of internal defects – microcracks and a microcavities. Performance of this condition by manufacture LTSC as at breakage composite preparation is not restored is especially important, and internal defects conduct to destruction of fibers and reduction of density of a transport current.

For reception of profiles of not round cross-section most often use technological process of repeated drawing. Key parameter at designing of process of the drawing, allowing to supervise break and to predict process of accumulation of defects, pressure of drawing $\sigma_{BO\pi}$ [2–4], which should have the minimum value.

In practical calculations for an estimation of influence of the form of section of a profile, channel geometry, conditions of a friction and deformation degree on size it is convenient $\sigma_{BO\pi}$ to use the formulas resulted in Ref. [2]. However, recommended formulas do not allow to define area of optimum angles of the channel die, providing the minimum value $\sigma_{BO\pi}$.

18.2 THEORETICAL BASES

For definition σ_{BOx} assumptions are accepted:
- Borders of a zone of deformation flat;
- Deformable material ideal plastic;
- Pressure are distributed in regular intervals on preparation section;
- Owing to small corners the die channel we accept equality of radial and normal contact pressure and $\sigma_r \approx \sigma_n$.
- The condition of plasticity looks like – $\sigma_x + \sigma_n = \sigma_s$ [2], where σ_x – longitudinal pressure; σ_s – resistance to deformation;
- On a contact surface normal and pressure tangents are distributed in regular intervals on profile perimeter;
- Forces of a friction on a contact surface are defined by the law – $\tau = f_n \sigma_n$, where, f_n – friction coefficient.

Pressure of drawing on an exit from the die channel фильеры we will define under the formula,

$$\sigma_{BO\pi} = \sigma_\kappa \cdot \phi = (\sigma_1 + 2\Delta\sigma) \cdot \phi \qquad (1)$$

where $\bar{\sigma}_\kappa = \bar{\sigma}_1 + 2\Delta\sigma$; σ_1 – pressure on an exit from a narrowed part of the die channel owing to metal deformation; $\Delta\sigma$ – an increment of longitudinal pressure for the account of additional shifts on an input and an exit of a narrowed part of the die channel; ϕ – the coefficient considering influence of the calibrating zone of the die channel.

18.2.1 CORRECT MANY-SIDED PROFILE

We accept that the die channel has the form of the correct truncated pyramid, and position of a side of the channel фильеры is defined by a corner α_m between an axis of the channel and an average line of a side. The scheme of a zone of deformation the six-sided profile in the die channel is shown on Fig. 18.3.

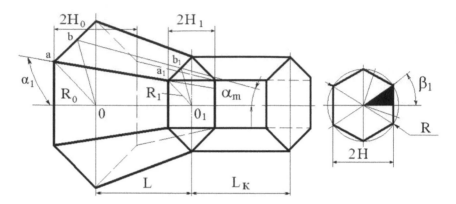

FIGURE 18.3   The scheme of a zone of deformation in the die channel.

R_0, R_1, R – radius of the described circle on an input and an exit of a zone of deformation, and current value of radius, accordingly; $2H_0$, $2H_1$ and $2H$ ($H = H_{0-1} \times \mathrm{tg}\alpha_m$) – section height on an entrance and an exit of a zone of deformation and its current value; L – length of a zone of deformation; α_m – tilt angle forming the die channel, a side passing through the middle.

Owing to symmetry for definition $\sigma_{BO\pi}$ it is enough to consider conditions of deformation of sector limited to planes aOb, $a_1O_1b_1$ aOO_1a_1 and bOO_1b_1 (Figs. 18.3 and 18.4).

It is similarly possible to allocate sector for any correct many-sided profile with any number of sides. The central corner of sector is calculated under the formula,

$$\beta_1 = \frac{\pi}{n},$$

where n – quantity of sides.

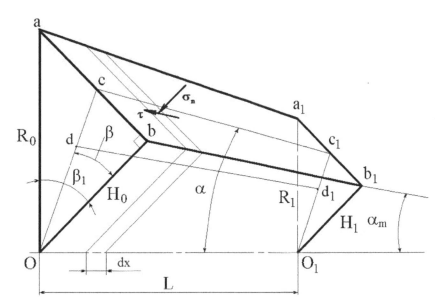

FIGURE 18.4   The scheme of sector of a zone of deformation.

Let's write down the differential equation of balance of an element of volume of a zone of deformation in length dx (Fig. 18.4)

$$-\sigma_{\delta} H^2 + (\sigma_x + d\sigma_x)(H + dH)^2 - 2f_n\sigma_n H\cos\alpha_m\, dx - 2\sigma_n H\sin\alpha_m\, dx = 0. \quad (2)$$

We accept that on the side aOb tension $\sigma_{x|x=0} = \sigma_q$, and on the side $a_1 O_1 b_{1}$. $\sigma_{x|x=L} = \sigma_1$. Considering that $dH / dx = -tg\alpha_m$, from the Eq. (2) we will receive,

$$\bar{\sigma}_1 = (1 + f_n \, ctg\alpha_m)(1 - \lambda^{-a}) + \bar{\sigma}_q \lambda^{-a}, \quad (3)$$

where $\bar{\sigma}_1 = \sigma_1 / \sigma_s; \bar{\sigma}_q = \sigma_q / \sigma_s; a = tg\alpha_m / f_n; \lambda = F_0 / F_1; F_0$ and F_1 – the area of section of a profile before and after deformation; $tg\alpha_m = (H_0 / L)(1 - 1/\sqrt{\lambda})$.

For definition $\Delta\sigma$, we use a method of balance of power.

Power of forces of shift on border aOb the die channel, being surface of rupture of speeds, is defined [5–7].

$$W = \int_S \tau_s |\Delta V_{II}|\, dS, \quad (4)$$

where $- \tau_s = \sigma_s / \sqrt{3}$ resistance to shift; ΔV_{II} – change components of speed of particles of the metal, a parallel plane aOb.

The angle of slope of a line of a current dd_1, located in a plane cOO_1c_1 (Figs. 18.4 and 18.5), is defined.

$$\text{tg}\alpha' = \bar{r} \cdot \text{tg}\alpha; \quad \text{tg}\alpha = \text{tg}\alpha_m / \cos\beta , \tag{5}$$

where $\bar{r} = r_0' / R_0' = h_0 / H_0$ – relative radius of a line of a current; $0 \le \bar{r} \le 1$; $0 \le \beta \le \beta_1$.

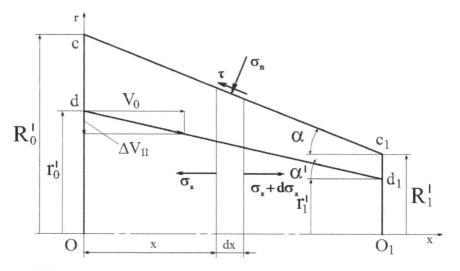

FIGURE 18.5 The scheme for definition $\Delta\sigma$.

Let's write down the equation of balance of capacities for a side aOb.

$$\Delta\sigma \, S \, V_0 = \int_S \tau_s |\Delta V_{II}| dS = \tau_s V_0 H_0^2 \, \text{tg}\alpha_m \int_0^1 \int_0^{\beta_1} \bar{r}^2 \frac{1}{\cos^3\beta} \, d\beta d\bar{r} = \frac{1}{6} \tau_s V_0 H_0^2 \, \text{tg}\alpha_m \, \text{tg}\beta_1 \left(\frac{1}{\cos\beta_1} + 1 \right). \tag{6}$$

From the Eq. (6) after transformations we will receive

$$\Delta\bar{\sigma} = \Delta\sigma / \sigma_s = 0,192 \cdot \text{tg}\alpha_m \cdot \text{tg}(\pi / n) / \text{tg}(\pi / 2n) . \tag{7}$$

Let's similarly define size $\Delta\bar{\sigma}$ for a side $a_1O_1b_1$.

Using Eqs. (1), (3) and (7), we will write down the equation,

$$\bar{\sigma}_\kappa = \bar{\sigma}_1 + 2\Delta\bar{\sigma} = (1+ f_n ctg\alpha_m)(1-\lambda^{-a}) + \bar{\sigma}_q \lambda^{-a} + 0,385 \cdot tg\alpha_m \cdot tg(\pi/n)/tg(\pi/2n), \qquad (8)$$

where $\bar{\sigma}_\kappa = \sigma_\kappa / \sigma_S$.

Let's simplify the Eq. (8). Accepting $\lambda^{-a} \approx 1 - a \ln \lambda$, we will receive,

$$\bar{\sigma}_\kappa = [1+ f_n ctg\alpha_m (1-\bar{\sigma}_q)] \cdot \ln\lambda + \bar{\sigma}_q + 0,385 \cdot tg\alpha_m \cdot tg(\pi/n)/tg(\pi/2n). \qquad (9)$$

For the account of influence of a calibrating zone of the die channel we will spend calculation on the greatest possible values of the normal pressure defined from a condition of plasticity [8].

$$\sigma_n = \sigma_s - \sigma_x$$

The differential equation of balance for elementary volume of preparation in a calibrating zone we will receive by analogy to the Eq. (2), believing $\alpha_m = 0$,

$$d\sigma_x F + 2f_n(\sigma_s - \sigma_x) P dx = 0, \qquad (10)$$

where F and P – the area and perimeter of section of sector in a calibrating zone, accordingly.

Size of pressure of drawing on an exit from the die channel $(x=L_\kappa)$ with the account of a boundary condition $\sigma_x|_{x=0} = \sigma_\hat{e}$ it will be defined.

$$\bar{\sigma}_{BO\pi} = 1 - (1-\bar{\sigma}_\kappa)\exp(-\delta_\kappa), \qquad (11)$$

where $\bar{\sigma}_{BO\pi} = \sigma_{BO\pi}/\sigma_S$; $\delta_\kappa = f_n(P/F)L_\kappa$; L_κ – length of a calibrating zone.

Let's simplify the Eq. (11). Accepting $\exp(-\delta_\kappa) \approx 1 - \delta_\kappa$ we will receive,

$$\bar{\sigma}_{BO\pi} = 1 - (1-\bar{\sigma}_\kappa)(1-\delta_\kappa). \qquad (12)$$

Difference of results of calculation between Eqs. (11) and (12) no more than 5% at $L_\kappa / H \leq 1$ and $f_n < 0.06$.

Factor δ_κ in Eqs. (13) and (14) simply to define, if the characteristic sizes of section of a polygon are known. Formulas for definition δ_κ are resulted in Table 18.1.

TABLE 18.1 Formulas for definition δ_κ.

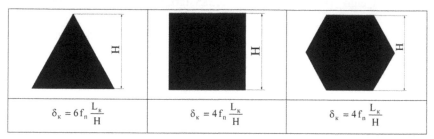

$\delta_\kappa = 6 f_n \dfrac{L_\kappa}{H}$	$\delta_\kappa = 4 f_n \dfrac{L_\kappa}{H}$	$\delta_\kappa = 4 f_n \dfrac{L_\kappa}{H}$

18.2.2 RECTANGULAR PROFILE

The scheme of a zone of deformation in a narrowed part of the die channel for a rectangular profile is shown on Fig. 18.6.

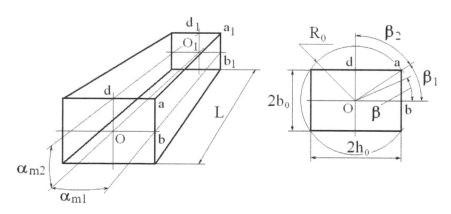

FIGURE 18.6   The scheme of a zone of deformation in the die channel.

Except accepted above assumptions also it is accepted:

- the die channel has the form of the correct truncated pyramid for which the parity of the parties of section remains to constants on length of a zone of deformation and is equal $-k=h_0/b_0$;
- position of sides of a profile of the die channel is defined by corners

α_{m1} (for a short side) and α_{m2} (for a long side) between an axis of the die channel and average lines of corresponding sides;

Let's define pressure of drawing σ_κ on an exit from a narrowed part of the die channel.

R_0, R_1, R – radius of a circle described round a profile on an input and an exit of a zone of deformation, and current value of radius, accordingly; $2h_0$ and $2b_0$ – height and width of section on an input; L – length of a zone of deformation; – an α_{m1} α_{m2} angle of slope of the forming die channel, passing through the middle of sides.

Owing to symmetry of a profile for definition of pressure σ_κ it is enough to consider conditions of deformation of sector limited to planes aOb, $a_1O_1b_1d_1$, $a_1O_1b_1$, bOO_1b_1.

For definition of pressure we use the schemes resulted in drawings 3 and 4, where instead of a corner it is necessary α_m to use a corner α_{m1}.

From geometrical parities we will receive $h = h_0 - x\,tg\alpha_{m1}$ and $dh/dx = -tg\alpha_{m1}$.

With the account of the accepted assumptions the differential equation of balance of an element of a zone of deformation dx (Figs. 18.2 and 18.3) and its decision are similar to the Eqs. (2) and (3).

$$\bar{\sigma}_1 = \sigma_1/\sigma_s = \left(1 + tg\alpha_{m1}/f_n\right)\left(1 - \lambda^{-a1}\right) + \bar{\sigma}_q\lambda^{-a1}, \tag{13}$$

where $a1 = f_n/tg\alpha_{m1}$; $tg\alpha_{m1} = (h_0/L)(1 - 1/\sqrt{\lambda})$.

The increment of pressure of drawing $\Delta\sigma$ we will define same as and for correct many-sided profiles.

From geometrical parities (Fig. 18.3) and the Eq. (6), we will receive expression for an angle of slope of a line of movement of particles of metal dd_1, located in planes Occ_1O_1 and capacity of forces of shift on the side Oab (Fig. 18.2).

$$tg\alpha' = \bar{r}\cdot tg\alpha; \quad tg\alpha = tg\alpha_{m1}/\cos\beta, \quad \beta = 0 \div \beta_1; \tag{14}$$

$$W = \frac{1}{6}\tau_s V_0 h_0^2 \, tg\alpha_{m1} \cdot tg\beta_1 \cdot \left(\frac{1}{\cos\beta_1}+1\right).$$ (15)

Considering, as $tg\beta_1 = b_0 / h_0 = 1/k$ and $tg(\beta_1 / 2) = \sin\beta_1 / (1+\cos\beta_1)$, from the Eq. (15), we will define.

$$\Delta\bar{\sigma}_1 = \Delta\sigma_1 / \sigma_s = \frac{1}{3\sqrt{3}} \cdot tg\alpha_{m1} \cdot [1+\sqrt{1+(1/k)^2}].$$ (16)

The increment of pressure of drawing $\Delta\sigma$ on the side $a_1 O_1 c_1$ also is defined by the Eq. (16).

Using received above expression, we will write down the equation for calculation of pressure of drawing of sector of the rectangular profile limited to a corner β_1.

$$\bar{\sigma}_{\kappa 1} = \bar{\sigma}_1 + 2\Delta\bar{\sigma}_1 = (1+tg\alpha_{m1}/f_n)(1-\lambda^{-a1}) + \bar{\sigma}_q \lambda^{-a1} + \frac{2}{3\sqrt{3}} tg\alpha_{m1} \, [1+\sqrt{1+(1/k)^2}].$$

Accepting $\lambda^{-a1} \approx 1 - a1 \cdot \ln\lambda$, we will simplify the received formula,

$$\bar{\sigma}_{\kappa 1} = \bar{\sigma}_1 + 2\Delta\bar{\sigma}_1 = \left[1+f_n ctg\alpha_{m1}(1-\bar{\sigma}_q)\right]\ln\lambda + \bar{\sigma}_q + \frac{2}{3\sqrt{3}} tg\alpha_{m1} \, [1+\sqrt{1+(1/k)^2}].$$ (17)

Let's similarly define pressure of drawing for the sector limited to a corner β_2.

$$\bar{\sigma}_{\kappa 2} = (1+tg\alpha_{m2}/f_n)(1-\lambda^{-a2}) + \bar{\sigma}_q \lambda^{-a2} + \frac{2}{3\sqrt{3}} \cdot tg\alpha_{m2} \cdot [1+\sqrt{1+(1/k)^2}],$$ (18)

where $\bar{\sigma}_2 = \sigma_2 / \sigma_s$; $a2 = f_n / tg\alpha_{m2}$; $tg\alpha_{m2} = (b_0 / L)(1-1/\sqrt{\lambda})$.

From geometrical parities it is had $tg\alpha_{m2} = tg\alpha_{m1}/k$. As a2=a1/k after transformations the Eq. (18) will become

$$\bar{\sigma}_{\kappa 2} = \bar{\sigma}_2 + 2\Delta\bar{\sigma}_2 = \left[1+kf_n ctg\alpha_{m1}(1-\bar{\sigma}_q)\right]\ln\lambda + \bar{\sigma}_q + 0,385\, tg\alpha_{m1} \, [1+\sqrt{1+(1/k)^2}]/k.$$ (19)

The average on section of a rectangular profile pressure of drawing is equal,

$$\bar{\sigma}_\kappa = (\bar{\sigma}_{\kappa 1} F_1 + \bar{\sigma}_{\kappa 2} F_2)/F,$$

where F – the area of section of a profile; F_1, F_2 – the total area of section of sectors limited to corners and β_1, β_2, accordingly.

As $F_1 = F_2 = F/2$, we will receive

$$\bar{\sigma}_K = [1 + (1+k)f_n \, ctg\alpha_{m1}(1-\bar{\sigma}_q)/2]\ln\lambda + \bar{\sigma}_q + \frac{1}{3\sqrt{3}}tg\alpha_{m1}[1+(1/k)(1+2\sqrt{1+(1/k)^2})] \, . \quad (20)$$

18.3   RESULTS AND THEIR DISCUSSION

The Eq. (8) confirms known experimental data that for the account of additional shifts pressure of drawing of a many-sided profile more than equal round profile. For comparison in Table 18.2, the data of calculation for $\bar{\sigma}_K$ different kinds of many-sided profiles is cited $\bar{\sigma}_K$ ($\alpha_m = 10^0$; $f_n = 0,1$; $\lambda = 1,2$ $\bar{\sigma}_q = 0$; $\bar{\sigma}_q = 0$) expression for the third composed in the Eq. (7).

TABLE 18.2   Settlement values $\bar{\sigma}_K$ and $2\Lambda\bar{\sigma}$.

n	3	4	5	6	∞ (circle)
$\bar{\sigma}_K$	0.475	0.435	0.423	0.418	0.407
$2\Lambda\bar{\sigma}$	$1,115 \cdot tg\alpha_m$	$0,929 \cdot tg\alpha_m$	$0,861 \cdot tg\alpha_m$	$0,829 \cdot tg\alpha_m$	$0,77 \cdot tg\alpha_m$

Using the Eq. (9), we will define the value of a corner α_m providing the minimum value of pressure of drawing from a condition,

$$\frac{d\bar{\sigma}_K}{dtg\alpha_m} = -\frac{f_n}{tg^2\alpha_m}(1-\bar{\sigma}_q)\ln\lambda + 0,385 \cdot tg(\pi/n)/tg(\pi/2n) = 0 \, .$$

The decision of this equation,

$$\alpha_m = \text{arctg}\left(1,612 \cdot \sqrt{f_n \cdot \ln\lambda \cdot (1-\bar{\sigma}_q) \cdot tg(\pi/2n)/tg(\pi/n)}\right) . \quad (21)$$

In Table 18.3, optimum values of a corner for the α_m correct many-sided profiles, calculated under the Eq. (21) are resulted α_m .

TABLE 18.3   Die channel optimum angle ($f_n = 0,06$ $\lambda = 1,15$; $\overline{\sigma}_q = 0$).

n	3	4	6	Circle ($n = \infty$)
α_m, degree	4.87	5.43	5.74	5.96

At use of the standard drawing tool value can α_m differ from the optimum. In this case optimum value is defined λ from the Eq. (10),

$$\lambda = \exp\left[0,385 \cdot \frac{tg(\pi/n)}{tg(\pi/2n)} \cdot \frac{tg^2\alpha_m}{f_n \cdot (1-\overline{\sigma}_q)} \right].\tag{22}$$

In Table 18.4, results of calculation of relative pressure of drawing of rectangular profiles for different values k are resulted. It is visible that with increase in a deviation of the form of a profile from square the pressure size increases $\overline{\sigma}_\kappa$, and the relative contribution of additional shifts decreases.

TABLE 18.4   Pressure of drawing of rectangular profiles ($\alpha_{m1} = 10^0$; $f_n = 0,1$ $\lambda = 1,2$; $\overline{\sigma}_q = 0$).

k	1	2	3	4
$\overline{\sigma}_K$	0.450	0.464	0.506	0.553
$2\Delta\overline{\sigma}$	0.164	0.127	0.117	0.112

Using the Eq. (22), we will define a corner α_{m1} providing the minimum value of pressure of drawing from a condition,

$$\frac{d\overline{\sigma}_{BO\pi}}{dtg\alpha_{m1}} = -\frac{f_n}{2\,tg^2\alpha_{m1}}(1+k)(1-\overline{\sigma}_q)\ln\lambda + \frac{1}{3\sqrt{3}} \cdot [1+(1/k)(1+2\sqrt{1+k^2})] = 0$$

.

The decision of this equation,

$$\alpha_{m1} = arctg\left(1,612 \cdot \sqrt{\frac{f_n \ln\lambda(1+k)(1-\overline{\sigma}_q)}{1+(1/k)(1+2\sqrt{1+k^2})}} \right).\tag{23}$$

In Table 18.5, the optimum values of a corner α_{m1} calculated under the Eq. (23), and values of a corner corresponding to them α_{m2}.

TABLE 18.5   Optimum values of corners of the die channel ($f_n = 0,1$; $\lambda = 1,2$; $\overline{\sigma}_q = 0$).

k	1	2	3	4
α_{m1}, degree	7.97	11.07	13.21	14.97
α_{m2}, degree	7.97	5.57	4.47	3.83

The analysis of results of calculation shows that the corner α_{m2} can be defined approximately under the formula (an error no more than 3%) – $\alpha_{m2} = \alpha_{m1}/k$.

The Eq. (23) allows to solve a return problem: for the set corners of the die channel to define optimum value λ

$$\lambda = \exp\left[0,385 \cdot \frac{1+(1/k)(1+2\sqrt{1+k^2})}{1+k} \cdot \frac{tg^2\alpha_{m1}}{f_n \cdot (1-\overline{\sigma}_q)}\right]. \qquad (24)$$

Influence of a calibrating zone is considered the same as and for many-sided profiles under Eqs. (13) or (14). Values for $\overline{\sigma}_K$ different rectangular profiles are resulted in Table 18.6.

TABLE 18.6   Values for $\overline{\sigma}_K$ different rectangular profiles

k	1	2	3	4
δ_K	$2L_K/b_1=2L_K/h_1$	$1.5L_K/b_1=3L_K/h_1$	$1.33L_K/b1=4L_K/h_1$	$1.25L_K/b_1=5L_K/h_1$

18.4   CONCLUSION

For continuous profiles of not round cross-section section the parities allowing are received: to calculate pressure of drawing; to define a corner of the channel of the drawing tool, providing the best power parameters of process of drawing; to calculate the best value of degree of deformation at use of the standard tool.

KEYWORDS

- composite
- drawing
- plastic deformation
- superconductor

REFERENCES

1. A.K. Shikov, etc. Development of superconductors for magnetic system of ITER in Russia [Razrabotka cverhprovodnikov dlya magnitnoy sistemy ITER v Rossii]. *Izvestiya vuzov. Tsvetnaya metallurgiya – News of higher education institutions. Nonferrous metallurgy*, 2003. No.1. p. 36–43.
2. Perlin I.L. Ermanok M. Z. Drawing theory [*Teoriya volocheniya*]. Moscow, Metallurgy, 1991. 456 p.
3. Kolmogorov V. L. Tension. Deformations. Destruction [*Napryazheniya. Deformatsii. Razrushenie*]. Moscow, Metallurgy, 1990. 229 p.
4. Kolmogorov V. L. Migachev B. A. Burdukovsky Century of. To a question of creation of the generalized phenomenological model of destruction at plastic deformation [*K voprosu postroeniya obobschonnoy fenomenologicheskoy modeli razrusheniya pri plasticheskoy deformatsii*]. *Metalli – Metals*, 1995. No.6. P.132–141.
5. Bekofen W. Deformation processing. Massachusetts, California, 1992. 288 p. 288 p.
6. Johnson W., Mellor P. The plasticity theory for engineers [*Teoriya plastichnosti dlya ingenerov*]. Moscow, Mechanical engineering, 1999.
7. Johnson W., Kudo H. Mechanics of process of expression of metal [*Mekhanika protsessov vidavlivaniya metalov*]. Moscow, Metallurgy, 1995. 174 p.
8. Emelyanenko P. T. Alshevsky L.E. *Stal – Steel*, 1999. No.10. p. 904.

CHAPTER 19

SYNTHESIS, STRUCTURE AND PROTECTIVE PROPERTIES OF TETRA SODIUM NITRILO-TRIS-METHYLENE PHOSPHONATE ZINCATE TRIDECAHYDRATE $Na_4 [N(CH_2 PO_3)_3 Zn] \times 13H_2O$

F. F. CHAUSOV, N. V. SOMOV, E. A. NAIMUSHINA, and I. N. SHABANOVA

CONTENTS

ABSTRACT

The authors have synthesized a coordination compound, which has the properties of corrosion inhibitor and studied its structure by the X-ray scattering method, X-ray photoelectron spectroscopy and Raman spectroscopy. The compound obtained has a chelate structure, which is stable during storing and provides high degree of steel protection against corrosion in neutral aqueous media. The crystal lattice belongs to triclinic system, the spatial group is $P\bar{1}$, Z=2, a = 11.2208(2) Å, B = 11.2666(3) Å, c = 12.3286(3) Å, α = 108.455(2)°, β = 97.168(2)°, γ = 117.103(2)°. The complex is chelate; the internal coordination sphere includes three five-membered cycles N–C–P–O–Zn with the common bond Zn–N. Zinc is coordinated in the distorted trigonal-bipyramidal configuration, which includes an oxygen atom of the neighboring molecule of the complex.

19.1 INTRODUCTION

Inhibitors of corrosion and salt-formation based on the complexes of organophosphonic acids with d-metals, mainly with zinc, are rather widespread [1–4]. However, the preparation of the inhibitors of this particular class with a prespecified structure and high-repeatability anticorrosive efficiency is very difficult because organophosphonic acids similar to other oxygen compounds of phosphor tend to the formation of spatial polymer structures [5]. Therefore, known marketable products contain particles of an inhibitor with nonstable structure and demonstrate changing inhibiting properties.

In known products [6–9], zinc ions are coordinated with oxygen atoms at the tops of a distorted octahedron and nitrogen atoms do not participate in coordinating a zinc ion. These inhibitors have certain drawbacks, namely, they tend to the formation of polymer structures and as a result, they are poorly dissolved in water and they have insufficiently high thermodynamic stability during storage and usage.

19.2   EXPERIMENT

Small portions of 0.1 mol of zinc oxide were added into the solution of 0.45 M sodium hydroxide in distilled water, which was heated to 70–80°C, at continual stirring. After complete dissolution, 0.1 M solution of recrystallized nitrilo-tris-methylene phosphonic acid was put in by drops at continual stirring. The mixture was stirred for 2 h at 70–80°C. The solution obtained was filtered, evaporated and slowly cooled down to room temperature. Settled crystals were separated on the filter.

The X-ray investigation was conducted on an automated four-circle diffractometer Oxford Diffraction Gemini S equipped with a CCD-detector Sapphire III at room temperature with the use of MoKα-radiation. From a monocrystalline sample, a polished spherule was made with an average radius of 0.16 mm. The primary fragment of the structure was found by a direct method. The positions of the other atoms (including hydrogen) were found in the differential synthesis of electron density. The structure was defined more exactly in an isotropic approximation for hydrogen atoms and in an anisotropic approximation for the other atoms. The results of the X-ray study were deposited in the **Cambridge Structural Database,** CCDC № 919565.

The X-ray photoelectron spectra were taken on an electron magnetic spectrometer with double focusing by magnetic field [10] at the excitation by AlKα-radiation. The temperature of the beginning of decomposition was measured in situ by a decrease in the contrast of the carbon C1s-spectrum on the same device.

The Raman spectra were obtained on a Raman-scattering microspectrometer "Centaur U HR" with the use of laser excitation with the wavelength of 532 nm.

Corrosion tests were conducted according to GOST 9.502–82 in static conditions in a neutral aqueous medium #2 according with the above GOST at natural aeration; the exposure was 10 days.

19.3   RESULTS AND DISCUSSION

Figure 19.1 shows a fragment of the crystal structure $Na_4[N(CH_2PO_3)_3Zn] \times 13H_2O$ in coordination polyhedrons and Fig. 19.2 presents the structure of the complex anion $[N(CH_2PO_3)_3Zn]^{4-}$.

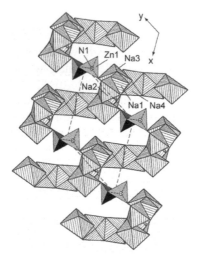

FIGURE 19.1   The structure $Na_4 [N(CH_2 PO_3)_3 Zn] \times 13H_2O$ in the coordination polyhedrons of zinc and sodium.

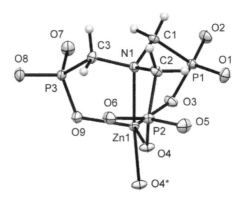

FIGURE 19.2   The structure of the complex anion $[N(CH_2 PO_3)_3 Zn]^{4-}$

   The nitrogen atom is deprotonated and it participates in the formation of the donor-acceptor bond with the atom of zinc which is coordinated in the configuration of the distorted trigonal bipyramid characteristic of the sp^3 d-hybridized state of the valence electron states; the distortion is due to the asymmetry of the coordination environment of the zinc atom and the formation of the bonds of Zn1–O4* and Zn1*–O4 with the complex coordinated in the symmetrically equivalent position $-x, -y, -z$ (the corresponding positions of the atoms are designated by asterix (*)). The plane configuration of the parallelogram Zn–O–Zn–O is explained by the presence of the center of symmetry. The atoms of N, Zn, O and P of the five-membered chelate cycle are situated in a coplanar manner; the atom C is "broken out" from the plane of the cycle. The coplanarity is due to the formation of the common binding electron subsystem of the atoms, N–Zn–O–P.

TABLE 19.1 Interatomic distances d and valence angles ω in the structure $Na_4[N(CH_2PO_3)_3Zn] \times 13H_2O$.

Bond	d, Å	Bond	d, Å	Angle	w, degree	Angle	w, degree	Angle	w, degree
Zn1–N1	2.2548(7)	P1–O2	1.5131(7)	N1–Zn1–O3	86.57(3)	O7–P3–O8	112.41(4)	O9–P3–C3	103.80(4)
Zn1–O9	1.9882(7)	P2–O6	1.5147(7)	N1–Zn1–O4	82.50(3)	O1–P1–C1	107.79(4)	O1–P1–O3	110.80(4)
Zn1–O4*	2.0151(6)	P3–O7	1.5162(7)	N1–Zn1–O9	87.11(3)	O2–P1–C1	108.38(4)	O2–P1–O3	112.24(4)
Zn1–O3	2.0184(7)	P1–O1	1.5193(7)	N1–Zn1–O4*	163.29(3)	O3–P1–C1	103.71(4)	O6–P2–O4	111.64(4)
Zn1–O4	2.0876(7)	P3–O9	1.5416(7)	O3–Zn1–O4	128.82(3)	O4–P2–C2	101.62(4)	O5–P2–O4	112.87(4)
P2–C2	1.8204(9)	P1–O3	1.5472(7)	O3–Zn1–O9	114.49(3)	O5–P2–C2	106.85(4)	O7–P3–O9	111.79(4)
P3–C3	1.8214(9)	P2–O4	1.5567(7)	O3–Zn1–O4*	102.68(3)	O6–P2–C2	109.90(4)	O8–P3–O9	112.09(4)
P1–C1	1.8217(9)			O4–Zn1–O9	114.66(3)	O7–P3–C3	109.26(4)	O1–P1–O2	113.34(4)
P2–O5	1.5108(7)			O4–Zn1–O4*	80.86(3)	O8–P3–C3	107.54(4)	O6–P2–O5	113.19(4)
P3–O8	1.5127(7)			O9–Zn1–O4*	101.27(3)				

*Symmetrically equivalent position: *–x, –y, –z.

The main data on the interatomic distances and valence angles are presented in Table 19.1.

The P2p spectrum of phosphorus of the obtained compound Fig. 19.3 contains two components with maxima at the binding energy 131.2 and

132.8 eV. The component with the maximum at 131.2 eV corresponds to the participation of a phosphorus atom in a single bond with an oxygen atom, and the component with the maximum at 132.8 eV corresponds to the participation of a phosphorous atom in a multiple bond with an oxygen atom [11]. The intensity ratio of these components is 1:2; it means that in the substance obtained there is a single bond between each phosphorous atom and an oxygen atom participating in the coordination bond with a zinc atom and there are two similar multiple bonds between each phosphorous atom and oxygen atoms which carry the same partial negative charges. It is in agreement with the data of the X-ray scattering analysis (Table 19.1), according to which the lengths of the P-O(Zn) bonds are larger on average by 0.0034 nm than the lengths of the $P-O^{\delta-}$ bonds which are almost the same (the difference in their lengths does not exceed 0,0006 nm). Figure 19.4 presents the correlation plot of the dependence of the lengths of the bonds on their bond order; the lengths of the single P–O(H) and double P=O bonds are taken according to the literature data [12], 0.1534 ± 0.0003 and 0.1477 ± 0.0003 nm, respectively. Linear interpolation gives the bond order of about 1.4 for the $P-O^{\delta-}$ bonds (at the average bond length 0.15145 ± 0.00007 nm).

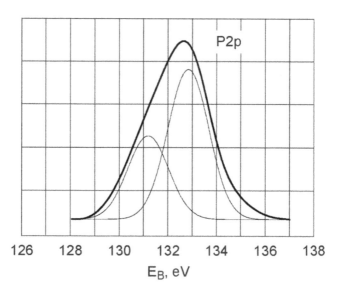

FIGURE 19.3   The P2p X-ray photoelectron spectrum of Na$_4$[N(CH$_2$PO$_3$)$_3$Zn]×13H$_2$O

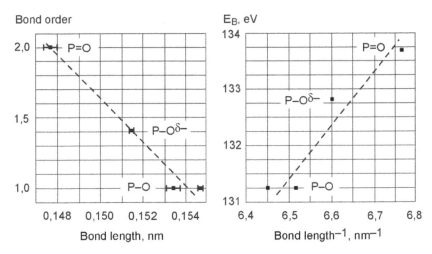

FIGURE 19.4   The bond order vs. P-O bond length correlation plot.

The above-described structure is also confirmed by the molecular-vibrational spectrum of the compound obtained (Table 19.2). The theoretical spectrum for the obtained structure has been conducted on a computer with the use of the program HyperChem 8.0.7. There is good agreement between the experimental and calculation data.

TABLE 19.2   The Raman spectrum of a single crystal $Na_4[N(CH_2PO_3)_3Zn] \times 13H_2O$.

Frequency, cm^{-1}		Correspondence
Experiment	**Calculation**	
100...300	100...300	Twist-vibrations of outer coordination sphere
334...400	307...418	d-vibrations of O–P–O
478	465	d-vibrations of Zn O P
571	500...590	d-vibrations of (Zn)O–P–O
620...700	620...700	Vibrations of outer coordination sphere
761	720...778	d-vibrations of N–C–P
846	824	Vibrations of outer coordination sphere
878	882...895	d-variations of Zn–O–P

TABLE 19.2 *(Continued)*

Frequency, cm⁻¹		Correspondence
Experiment	Calculation	
918	920	Vibrations of outer coordination sphere
1003	999…1002	Symmetric twist- vibrations of cycles Zn–N–C–P–O(Zn)
1080…1160	1080…1160	Raman modes of outer coordination sphere and cycles Zn–N–C–P–O(Zn)
1184	1176	Twist- vibrations of cycle Zn–O–Zn–O(Zn)
1243	1230…1235	n-vibrations of $(O_3)Zn–N(C_3)$
1399	1393	Plane vibrations of cycle Zn–O–Zn–O(Zn)
1435	1427…1428	Antisymmetric twist-vibrations of cycles Zn–N–C–P–O(Zn)
2824	2802…2830	$H_2O×××Na(O–P)$
2827…2988	2911…3022	$H_2O×××Na(nH_2O)$
3200…3500	3200…3500	$H_2O××× H_2O$

TABLE 19.3 The comparison of the efficiency and thermal stability of the known and obtained inhibitors.

Inhibitor	Degree of the corrosion protection, %	Efficiency of the salt-formation inhibition, %	Temperature of the beginning of decomposition, °C
"Ectoscale-450-2"	91	97	218
Product obtained	93	99	250

The comparison of the inhibition efficiency and the temperature of the beginning of the decomposition of the known inhibitor "Ectoscale-450-2", patent [9], and compound obtained is displayed in Table 19.3. The better protective properties of the obtained compound should be ascribed to its regular and stable structure; the thermal stability demonstrated by the high temperature of the beginning of the decomposition is due to the formation of the strong Zn-N bond with the closure of three five-membered chelate cycles.

19.4 CONCLUSIONS

A new inhibitor of steel corrosion in neutral media is synthesized, which has a chelate structure containing the Zn-N bond and in contrast to the prototype having better efficiency and stability.

The zinc atom is in the distorted trigonal-bipyramidal environment of the N- and O-atoms of the molecule of the nitrilo-tris-methylene phosphonic acid. The complex anions are dimerized by the Zn-O bonds. The outer coordination sphere contains sodium ions in the distorted trigonal-bipyramidal and octahedral environment of oxygen atoms and the molecules of solvation water.

The phosphonate groups contain the P-O(Zn) single bond and two approximately 1.4-multiple bonds $P–O^{\delta-}$.

KEYWORDS

- **Chelate complexes**
- **coordination compounds**
- **corrosion inhibitors**
- **Nitrilo-tris-methylene phosponates**
- **Zinc**

REFERENCES

1. Balaban-Irmenin Yu. V., Rudakova G. Ya., Markovich L. M. The application of scale-control additives in the energy of low parameters. M. : "Novosti teplosnabzheniya. " 2011. 208 p.
2. Balaban-Irmenin Yu. V., Lipovskikh V. M., Rubashov A. M. Protection against internal corrosion of heat network pipelines. 2 izd. M. : "Novosti teplosnabzheniya. " 2008. 288 p.
3. Chausov F. F. New effective method for the protection of heat-transfer equipment against salt-formation Tyazholoye mashinostroeniye. 2007. №9. P. 5–8.
4. Chausov F. F. Effective method for protecting steel equipment of network against corrosion Ekologia i promyshlennost Rossii. 2009. №2. P. 8–12.

5. Dyatlova N. M., Tyomkina V. Ya., Popov K. I. Complexons and complexonate of metals. M. : Khimia, 1988, 544 p.

6. Demadis K. D., Mantzaridis C., Raptis R. G., Mezei G. Metal-Organotetraphosphonate Inorganic-Organic Hybrids: Crystal Structure and Anticorrosion Effects of Zinc Hexa methylenediaminetetrakis(methylenephosphonate) on Carbon Steels Inorganic Chemistry, Vol. 44, No. 13 (2005). P. 4469–4471.

7. Demadis K. D., Barouda E., Zhao H., Raptis R. G. Structural architectures of charge-assisted, hydrogen-bonded, 2D-layered amine-tetraphosphonate and zinc-tetraphosphonate ionic materials Polyhedron, 28 (2009). P. 3361–3367.

8. Demadis K. D., Katarachia S. D., Koutmos M. Crystal growth and characterization of zinc-(amino-tris-(methylenephosphonate)) organic-inorganic hybrid networks and their inhibiting effect on metallic corrosion Inorganic Chemistry Communications, 8 (2005). P. 254–258.

9. Patent RF №2115631. Composition for inhibiting salt-formation and corrosion / Kovalchuk A. P., Ivanova N. A. MPK C 02 F 5/14, published 20. 07. 1998.

10. Trapeznikov V. A., Shabanova I. N., Terebova N. S., Murin A. V., Naimushina E. A. Investigation of the electronic structure of the transition metal-based system at varying concentration, temperature and pressure. M. Izhevsk: Udmurt State University, 2011. 216 p.

11. Shabanova I. N., Chausov F. F., Naimushina E. A., Kazantseva I. S. Application of the X-ray photoelectron spectroscopy method for investigating the molecular structure of the corrosion inhibitor complex of 1-hydroxietylidene diphosphonic acid with zinc Zhurnal strukturnoi khimii. 2011. Vol. 52. Supplement. C. S113–S118.

12. Dali J. J., Wheatley P. J. The crystal and molecular structure on nitrilotrimethylene triphosphonic acid Journal of Chemical Society (A), 1997, Vol. 24, P. 212–221.

PART VI

NANOSYSTEMS AND NANOSTRUCTURES
IN NANOBIOLOGY AND NANOMEDICINE

CHAPTER 20

INFLUENCE OF GEOMAGNETIC FIELD VARIATION AND FE CONCENTRATION IN THE FEED ON FEED MINERALIZATION IN THE ONTHOGENESIS OF *APIS MELLIFERA* (L.)

G. V. LOMAEV, N. V. BONDAREVA, and M. S. EMELYANOVA

CONTENTS

ABSTRACT

It has considered the influence of hypogeomagnetic fields and high content of iron in the diet on the process of mineralization in the ontogeny of bee *Apis melifera* L. It has established that mineralization begins at 8–9 day larval stage in gipermagnetic and hypomagnetic fields with the conditions of increased ingress of iron into the body, whereas it begins at 10–11 day after oviposition if content of iron is natural. Variations of Earth's magnetic field also influence on the process. The most significant manifestation of this magnetobiological effects observed in compensated geomagnetic field – ferromagnetic phase forms on 12 day.

20.1 INTRODUCTION

The presence of natural magnetic fields in the environment is a normal and essential factor for a rational human life and every living system. The absence of these fields or their deficiency may be as a factor, which launch processes of homeostatic regulation of functional systems so a factor, which lead to the growth of abnormalities in different stages of organism's development. It is detected behavioral and physiological responses of biological objects, which have different levels of organization to changes of the geomagnetic field: mutation rate increases, the formation of malignant tumors stimulates, the functions of central nervous system changes, the ability to orient interferes and appear environmental anomalies. The most high-sensitivity to magnetic fields (MF) was detected in migratory fish, sea turtles, birds and honey bees *Apis melifera* L [1, 2].

Discovered high magnetic sensitivity of bees, features of lifestyle of individuals and families in general, widespread in nature and the relative availability of these animals provide researchers great manipulative capabilities and allow to use these organisms for research of the magnetic field effect on biological objects.

Most of experiments allow to establish the magnetic sensitivity of bees, but subtle mechanisms of perception MF are still inexplicable. It has been proposed various hypotheses, which explain the high magnetic sensitivity of organisms. Today, preference is given to "magnetite hypothesis," which suggest that the basis of bacteria's magneto reception and most terrestrial organisms is the interaction of an external magnetic field with magnetite crystals – biogenic iron compound Fe_3O_4, which have the properties of ferrite.

Magnetite crystallizes in the spinel structure. The unit cell of spinel is a face-centered cube, 32 oxygen atoms form the skeleton, which are located in the voids of the iron cations. Some magnetic properties (e.g., residual magnetization, coercive force) depend on the size of particles that is the domain of state. The concept of domains is one of the mainest in magnetism. Magnetic domains are called the magnetic material inside which the magnetization is uniform and directed everywhere in the same way. The saturation magnetization of magnetite is maximum at $T = 0$ K (corresponding to the specific magnetic moment is equal to 90 * G cm_3/g) and turns to zero at the Curie temperature T_c, when the thermal energy of the KTC (k – Boltzmann constant) is equal to the energy of exchange interaction. Researches of the magnetic hysteresis of magnetite indicate that the normal residual magnetization in fields less than 500 occurs due to the translational displacement of the domain walls, and in large fields due to their irreversible turning. [3]

So, the magnetite crystals, which present in the body of magnetically sensitive animals and form as result of processes of iron bio mineralization, are some MF detector, which is based on the mechanism of interaction of the external magnetic field with the crystal.

The fact that the magnetic crystals may actually be responsible for animals' magneto reception demonstrated only for bacteria. It is necessary a few thousand cells, which contain magnetic for the perception of the local gradients of geomagnetic field with an accuracy of 0.1%. The entire receptor system may take 1 mm3 with a share of 10.4% magnetite.

Magnetite crystals were found in different taxonomic groups. And a variety of organisms – from bacteria to vertebrates – has ability biosynthesis of magnetite.

The mechanisms and dynamics of the magnetite formation during the ontogeny of bees are still inexplicable. The degree of biological control of the dynamics of ferromagnetic material formation also has not been researched. In particular, the it is unclear what the role in ontogeny of iron biomineralization played such factors, as change in the amount of MF and iron. Experiments, which aimed at the research of iron mineralization, when the amounts are increased and distorted geomagnetic field have never been carried out.

20.2 EXPERIMENTAL

In the research laboratory FGBOU "Izhevsk State Technical University named after MT Kalashnikov "conducted experiments on the effect of weak magnetic field on various biological objects. In particular, an experiment to study the mineralization process of iron in the ontogeny of bees Apis mellifer L. The objectives of this study were: to check the appearance of magnetite phase in the ontogeny of the bees, study of the effect of excessive intake of iron in the body at the beginning of mineralization of the ferromagnetic phase, the registration of the appearance of magnetite crystals ponderomotive method.

Ponderomotive method allows to determine the magnetic characteristics (in particular, the magnetic moment) of the specimen by measuring the force acting on the sample placed in a magnetic field. Sufficiently high sensitivity allows extensive use of the method for the study of weakly magnetic samples.

Test examples were the bees at early development stages selected from reference- and test bee colonies. Bees from the reference group №1 were developing under natural conditions. Bee groups №2 and №3 were kept under the conditions of increased ingress of iron into the body (excess of 5 and 10 times, respectively, as compared to the normal conditions) for two months. Bees from the group №4 were developing under hypogeomagnetic field, while the bees of groups №5, №6 and №7 – under hypergeomagnetic field (the field was, respectively, 2, 5 and 10 times more intensive than the terrestrial field).

20.3 RESEARCH TECHNIQUES

Appearance of ferromagnetic phase was recorded by means of ponderomotive method. Magnetic torque, acting on an example, is described by the following equation:

$$M_{вр} = \mu_0 \times H \times M \times \sin \alpha, \tag{1}$$

where μ_0 – magnetic constant; H – magnetic field intensity; M – magnetic moment of the example; α – angle between the vector direction of the magnetic moment of the example and the vector of the magnetic field.

Magnetic moment was excited by electromagnet. The field intensity between magnet poles reached 10^8 $A \times m^{-1}$.

Suspension torque is described by the formula

$$M_{кр} = \varphi \times g \times G \times J_p/l, \tag{2}$$

where φ – suspension twist angle; g – free fall acceleration (g = 9,8 $m \times s^2$); G – modulus of elasticity in shear, $kg \times mm^{-2}$; J_p – polar moment of inertia, mm^{-4}; l – suspension length, m. Polar moment of inertia J_p depends on suspension radius R: $J_p = \pi R^4/2$.

Wire diameter in the installation was chosen equal to 6×10^{-5} m. Thus, the polar moment of inertia calculated by the formula above makes 1272×10^{-18} m^{-4}, suspension length is 2.5 m.

According to the Eq. (2), the torque arising at suspension twist angle of 0.1 radian is equal to

$$M_{кр} = 2.1 \times 10^{-9} \; kg \times m^2 \times s^{-2}.$$

In equilibrium:

$$M_{вр} = M_{кр}, \tag{3}$$

$$\mu_0 \times H \times M \times \sin \alpha = M_{кр.} \tag{4}$$

$$M = M_{кр}/\mu_0 \times H \times \sin \alpha. \tag{5}$$

Maximum sensitivity of the experimental installation is reached at $\sin \alpha = 1$. Substituting in the equation (5) H = 10^8 $A \times m^{-1}$, $\alpha = 90°$ and $M_{кр} = 2.1 \times 10^{-9}$ $kg \times m^2 \times s^{-2}$, we find the installation threshold of sensitivity:

$$M_{порог} = 1.67 \times 10^{-11} \; A \times m^2.$$

As it follows from the calculations above, threshold of sensitivity of the ponderomotive method is comparable with the one of SQUID-magnetometer [3].

1) External magnetic field generation.

Magnetic field generator allowing field intensity variation within the bee ball represents three-dimensional Helmholtz coils, that is three pairs of mutually parallel coils (X, Y, Z) constructively combined into one unit ("cube") and connected to power supply (deviations of magnetic field in various points within the system of coils, as compared to the central point, are below 5%).

2) Change of iron concentration in the feed.

To keep the iron dose in the ration constant, it was necessary to restrict the flying activity of bees (in that way we can exclude the ingress of iron from the external environment), remove from the hive the earlier made reserves of honey and cerago (ambrosia) and change to artificial feed with specially selected concentration of iron.

Reference- and test families received daily a certain amount (500 mL) of syrup with known concentration of iron. The syrup consisted of sugar (50%), water (40 %) and milk (10%) [4].

The family №1 (the reference family) received daily 500 mL syrup without any special ferriferous additives. The family №2 received 500 mL syrup with $FeCl_3$ solution. Fe concentration in the solution was 5 times higher than the reference concentration. The family №3 used a syrup with $FeCl_3$ additives, where Fe concentration was almost 10 times higher that the reference one.

20.4   RESULTS AND DISCUSSION

Using ponderomotive method, the magnetic moments of adult bees and pupae from all groups were recorded (Table 20.1) and their ages were fixed as soon as the magnetic phase appeared (after the sensitivity threshold of the installation was exceeded, of course). We failed to register magnetic substance at the stage of uncapped brood (larvae and eggs). Exception were insects from groups №5 and №6, developed under the conditions of increased ingress of iron into the body. Presence of the magnetic moment were registered for that groups at late larva stage (8–9th day from the time of egg laying).

TABLE 20.1   Registration of ferromagnetic phase appearance in the onthogenesis of bees.

Stage	Age (days)	Family (group of bees)						
		ref.	Fe*5	Fe*10	H=0	H*2	H*5	H*10
		1	2	3	4	5	6	7
Imago		+	+	+	+	+	+	+
Pupa	13–21	+	+	+	+	+	+	+
	12	+	+	+	+	+	+	+
Prepupa	11	+	+	+	−	+	+	+
	10	+	+	+	−	+	+	+
Larva	9	−	+	+	−	−	−	−
	8	−	+	+	−	−	−	−
	4–7	−	−	−	−	−	−	−
Egg	1–3	−	−	−	−	−	−	−

   the absence of the ferromagnetic phase.

   registration of the ferromagnetic phase.

Experimental results indicate that bio-mineralization of iron and formation of magnetically ordered magnetite compound starts as early as at preimago stage of bee development. The conclusion agrees with the data of J. Gould [2], who determined, by means of SQUID-magnetometry, residual magnetic moments and saturation moments of elder pupae.

Similar experiments, made on monarch butterflies, allow to assume, that small but meaningful quantity of magnetic substance is synthesized at the transition stage from larva to the adult insect [1]. The obtained experimental data indicate that normally imago, pupae and prepupae of working bees have noticeable magnetic moments.

Mineralization is subjected to change under the conditions of excessive ingress of iron (families № 2 and №3). Thus, experiments reveal that insects at late larva stage (8–9 weeks after the egg laying) have noticeable magnetic moments. At first sight, it seems to be connected with high concentration of iron in larva tissue [4] and with the biological necessity of keeping the element homeostasis (for instance, by keeping it in the chemically inert form of Fe_3O_4). However, we observed no unambiguous relation between iron concentration in the insect body and the beginning of ferromagnetic phase mineralization. Under natural conditions, bees' ferromagnetic phase was registered 10–11 days after the egg laying. At bees of group №2 and №3 (developed under increased ingress of iron) the comparable iron concentration was reached as early as at the 5th day of development. However, ferromagnetic phase appeared only at the 8–9th day.

Processes of ferromagnetic phase formation in the bodies of bees exposed to hypergeomagnetic field (families №5, №6 and №7) undergo not noticeable change. It was observed that its formation, just as under natural conditions, starts at the prepupa stage (10–11th day after the egg laying), despite the fact that iron concentration in tissues becomes stable already at the 6–7th day of the development.

For bees developed under hypogeomagnetic field (family №4) the ferromagnetic phase could be registered starting from pupa stage only (12 days after the egg laying). That may result either from the absence (or small concentration) of ferromagnetic phase in the example or from the positional relationship of crystals in the tissues of bee giving zero total magnetic moment. We can't exclude the "directing effect" of magnetic field on the crystal growth, whereas in the absence of the field (its compensation) the crystals can grow disorderly and in all directions which results in zero total magnetic moment. However, magnetic moment can be registered at pupa- and imago stages.

20.5 CONSLUSIONS

1. Under natural conditions, the ferromagnetic phase in the onthogenesis of bees is recorded at the prepupa stage (10–11 days after the egg laying).

2. It is experimentally proved that mineralization process can be influenced by external magnetic fields and concentration of iron in the feed.
3. Under the conditions of increased ingress of iron, mineralization of ferromagnetic substance starts at larva stage (8–9th day).
4. Hypergoemagnetic fields do not influence noticeably on the beginning of magnetic substance formation in the onthogenesis of bees.
5. Ferromagnetic phase for bees developed under the conditions of compensated geomagnetic field was recorded at pupa stage (12 days after the egg laying).

KEYWORDS

- geomagnetic field
- hypergoemagnetic fields
- mineralization of biomagnetit
- the iron content in the feed

REFERENCES

1. Biogenic magnetite and magnetoreception. New in biomagnetism. B.2: translation from English. edited, *Kirschvink J. L., D. McFadden.* M.: Mir, 1989. 576 p.
2. Gould J. L., Kirschvink J. L., Deffeyes K. S. Bees have magnetic remanence. Science. 1978. Vol. 201. №. 4360. P. 1026–1028.
3. Lomaev G. V., Bondareva N. V. Magnetic parameters of bees Apis mellifera mellifera (L.), obtained SKVD-magnetometry. Biophysics. 2004. B.49. №. 6. C. 1118–1119.
4. Lomaev G. V., Bondareva N. V. The dynamics of accumulation of iron in the body of the bee and the products of its activity. Environmental issues and natural resources in the agricultural sector: materials All-Russian Scientific and Practical Conference (Izhevsk, 2003). M.: ANK, 2003. pp. 171–180.

CHAPTER 21

TRANSFORMATION OF HIGH-ENERGY BONDS IN ATP

G. A. KORABLEV, N. V. KHOKHRIAKOV, G. E. ZAIKOV, and YU. G. VASILIEV

CONTENTS

ABSTRACT

With the help of spatial-energy concept it is demonstrated that the formation and change of high-energy bonds in ATP take place at the functional transitions of valence-active orbitals of the system "phosphorus-oxygen."

These values of energy bonds are in accord with experimental and quantum-mechanical data.

21.1   SPATIAL-ENERGY PARAMETER

During the interaction of oppositely charged heterogeneous systems the certain compensation of volume energy of interacting structures takes place which leads to the decrease in the resulting energy (e.g., during the hybridization of atomic orbitals). But this is not the direct algebraic deduction of the corresponding energies. The comparison of multiple regularities of physical, chemical and biological processes allows assuming that in such and similar cases the principle of adding the reciprocals of volume energies or kinetic parameters of interacting structures is executed.

Lagrangian equation for the relative movement of the system of two interacting material points with the masses m_1 и m_2 in coordinate x is as follows:

$$\text{Mred } x'' = -\frac{\partial U}{\partial x}, \text{ where } \frac{1}{m_{red}} = \frac{1}{m_1} + \frac{1}{m_2} \qquad (1), (1a)$$

Where, U – mutual potential energy of material points; m_{red} – reduced mass. Herein $x'' = a$ (system acceleration).

For the elementary areas of interactions Δx we can accept: $\dfrac{\partial U}{\partial x} \approx \dfrac{\Delta U}{\Delta x}$.

Then: $m_{red} a \Delta x = -\Delta U;$ $\qquad \dfrac{1}{1/(a\Delta x)} \cdot \dfrac{1}{(1/m_1 + 1/m_2)} '' \Delta U$

or: $\dfrac{1}{1\!\!\big(\,m_1 a\Delta x\,\big) + 1\!\!\big(\,m_2 a\Delta x\,\big)} '' \Delta U$

Since the product $m_i a \Delta x$ by its physical sense equals the potential energy of each material point $(-\Delta U_i)$, then:

$$\frac{1}{\Delta U} \approx \frac{1}{\Delta U_1} + \frac{1}{\Delta U_2} \tag{2}$$

Thus the resulting energy characteristic of the system of two interacting material points is found by the principle of adding the reciprocals of initial energies of interacting subsystems.

"The electron with the mass m moving about the proton with the mass M is equivalent to the particle with the mass: $m_{red} = \dfrac{mM}{m+M}$ " [1].

Therefore, modifying the Eq. (2) we can assume that the energy of atom valence orbitals (responsible for interatomic interactions) can be calculated [2] by the principle of adding the reciprocals of some initial energy components based on the equations:

$$\frac{1}{q^2/r_i} + \frac{1}{W_i n_i} = \frac{1}{P_E} \tag{3}$$

or

$$\frac{1}{P_0} = \frac{1}{q^2} + \frac{1}{(Wrn)_i} ; \tag{4}$$

$$P_E = P_0/r_i \tag{5}$$

here: W_i – orbital energy of electrons [3]; r_i – orbital radius of i orbital [4]; $q = Z^*/n^*$ – by Refs. [5, 6], n_i – number of electrons of the given orbital, Z^* and n^* – nucleus effective charge and effective main quantum number, R – bond dimensional characteristics.

The value P_0 is called a spatial-energy parameter (SEP), and the value P_E – effective P–parameter (effective SEP). Effective SEP has a physical sense of some averaged energy of valence orbitals in the atom and is measured in energy units, for example, in electron-volts (eV).

The values of P_0 parameter are tabulated constants for electrons of the given atom orbital.

For SEP dimensionality:

$$[P_0]=[q^2]=[E]\cdot[r]=[h]\cdot[\upsilon]=\frac{kgm^3}{s^2}=\text{J m,}$$

where [E], [h] and [υ] – dimensionalities of energy, Planck's constant and velocity.

The introduction of P-parameter should be considered as further development of quasi-classical concepts with quantum-mechanical data on atom structure to obtain the criteria of phase-formation energy conditions. For the systems of similarly charged (e.g., orbitals in the given atom) homogeneous systems the principle of algebraic addition of such parameters is preserved:

$$\sum P_E = \sum (P_0/r_i); \tag{6}$$

$$\sum P_E = \frac{\sum P_0}{r} \tag{7}$$

or:

$$\sum P_0 = P_0' + P_0'' + P_0''' + \cdots; \tag{8}$$

$$r\sum P_E = \sum P_0 \tag{9}$$

Here P-parameters are summed up by all atom valence orbitals.

To calculate the values of P_E-parameter at the given distance from the nucleus either the atomic radius (R) or ionic radius (r_I) can be used instead of R depending on the bond type.

Let us briefly explain the reliability of such an approach. As the calculations demonstrated the values of P_E-parameters equal numerically (in the range of 2%) the total energy of valence electrons (U) by the atom statistic model. Using the known correlation between the electron density (β) and intraatomic potential by the atom statistic model [7], we can obtain the direct dependence of P_E-parameter on the electron density at the distance r_i from the nucleus. The rationality of such technique was proved by the calculation of electron density using wave functions by Clementi [8] and comparing it with the value of electron density calculated through the value of P_E-parameter.

The modules of maximum values of the radial part of Ψ-function were correlated with the values of P_0 parameter and the linear dependence between these values was found. Using some properties of wave function as applicable to P-parameter, the wave equation of P-parameter with the formal analogy with the equation of Ψ-function was obtained [9].

21.2 WAVE PROPERTIES OF P-PARAMETERS AND PRINCIPLES OF THEIR ADDITION

Since P-parameter has wave properties (similar to Ψ'-function), the regularities of the interference of the corresponding waves should be mainly fulfilled at structural interactions.

The interference minimum, weakening of oscillations (in antiphase) occurs if the difference of wave move (Δ) equals the odd number of semi-waves:

$$\Delta = (2n+1)\frac{\lambda}{2} = \lambda\left(n+\frac{1}{2}\right), \text{ where n = 0, 1, 2, 3, ...} \tag{10}$$

As applicable to P-parameters this rule means that the interaction minimum occurs if P-parameters of interacting structures are also "in antiphase" – either oppositely charged or heterogeneous atoms (e.g., during the formation of valence-active radicals CH, CH_2, CH_3, NO_2 ..., etc.) are interacting.

In this case P-parameters are summed by the principle of adding the reciprocals of P-parameters – Eqs. (3) and (4).

The difference of wave move (Δ) for P-parameters can be evaluated via their relative value $\left(\gamma = \dfrac{P_2}{P_1}\right)$ of relative difference of P-parameters (coefficient α), which at the interaction minimum produce an odd number:

$$\gamma = \frac{P_2}{P_1} = \left(n+\frac{1}{2}\right) = \frac{3}{2}; \frac{5}{2}\ldots\ldots \text{ When } n = 0 \text{ (main state) } \frac{P_2}{P_1} = \frac{1}{2} \tag{11}$$

It should be pointed out that for stationary levels of one-dimensional harmonic oscillator the energy of these levels $\varepsilon=hv(n+\dfrac{1}{2})$, therefore, in quantum oscillator, in contrast to the classical one, the least possible energy value does not equal zero.

In this model the interaction minimum does not provide zero energy corresponding to the principle of adding reciprocals of P-parameters – Eqs. (3) and (4).

The interference maximum, strengthening of oscillations (in phase) occurs if the difference of wave move equals the even number of semiwaves:

$$\Delta=2n\frac{\lambda}{2}=\lambda n \text{ or } \Delta=\lambda(n+1).$$

As applicable to P-parameters the maximum interaction intensification in the phase corresponds to the interactions of similarly charged systems or systems homogeneous by their properties and functions (e.g., between the fragments or blocks of complex inorganic structures, such as CH_2 and NNO_2 in octogene).

And then:

$$\gamma=\frac{P_2}{P_1}=(n+1) \tag{12}$$

By the analogy, for "degenerated" systems (with similar values of functions) of two-dimensional harmonic oscillator the energy of stationary states:

$$\varepsilon=hv(n+1)$$

By this model the interaction maximum corresponds to the principle of algebraic addition of P-parameters – equations (6–8). When $n=0$ (main state) we have $P_2=P_1$, or: the interaction maximum of structures occurs if their P-parameters are equal. This concept was used [2] as the main condition for isomorphic replacements and formation of stable systems.

21.3   EQUILIBRIUM-EXCHANGE SPATIAL-ENERGY INTERACTIONS

During the formation of solid solutions and in other structural equilibrium-exchange interactions the unified electron density should be established in the contact spots between atoms-components. This process is accompanied by the redistribution of electron density between valence areas of both particles and transition of a part of electrons from some external spheres into the neighboring ones.

It is obvious that with the proximity of electron densities in free atoms-components the transition processes between the boundary atoms of particles will be minimal thus contributing to the formation of a new structure. Thus the task of evaluating the degree of such structural interactions in many cases comes down to comparative assessment of electron density of valence electrons in free atoms (on the averaged orbitals) participating in the process.

Therefore the maximum total solubility evaluated via the structural interaction coefficient α is defined by the condition of minimal value of coefficient α, which represents the relative difference of effective energies of external orbitals of interacting subsystems:

$$\alpha = \frac{P'_o/r_i' - P''_o/r_i''}{(P'_o/r_i' + P''_o/r_i'')/2} 100\% \tag{13},$$

or

$$\alpha = \frac{P'_S - P''_S}{P'_S + P''_S} 200\% \tag{14}$$

where P_S – structural parameter is found by the equation:

$$\frac{1}{P_S} = \frac{1}{N_1 P'_E} + \frac{1}{N_2 P''_E} + \dots \tag{15}$$

here N_1 and N_2 – number of homogeneous atoms in subsystems.

The nomogram of the dependence of structural interaction degree (ρ) upon the coefficient α, the same for the wide range of structures, was prepared by the data obtained (the figure is not available).

Isomorphism as a phenomenon is usually considered as applicable to crystalline structures. But obviously the similar processes can also take place between molecular compounds where the bond energies can be assessed via the relative difference of electron densities of valence orbitals of interacting atoms. Therefore the molecular electronegativity is rather easily calculated via the values of corresponding P-parameters.

In complex organic structures the main role in intermolecular and intramolecular interactions can be played by separate "blocks" or fragments considered as "active" areas of the structures. Therefore it is necessary to identify these fragments and evaluate their spatial-energy parameters. Based on wave properties of P-parameter, the total P-parameter of each element should be found following the principle of adding the reciprocals of initial P-parameters of all the atoms. The resulting P-parameter of the fragment block or all the structure is calculated following the rule of algebraic addition of P-parameters of their constituent fragments.

Apparently, spatial-energy exchange interactions (SEI) based on leveling the electron densities of valence orbitals of atoms-components have in nature the same universal value as purely electrostatic coulomb interactions and complement each other. Isomorphism known from the time of E. Mitscherlich (1820) and D.I. Mendeleev (1856) is only a special demonstration of this general natural phenomenon.

The quantitative side of evaluating the isomorphic replacements both in complex and simple systems rationally fits into P-parameter methodology. More complicated is the problem of evaluating the degree of structural SEI for molecular structures, including organic ones. Such structures and their fragments are often not completely isomorphic to each other. Nevertheless, SEI is going on between them and its degree can be evaluated either semiquantitatively numerically or qualitatively. By the degree of isomorphic similarity all the systems can be divided into three types:

I Systems mainly isomorphic to each other – systems with approximately the same number of *heterogeneous* atoms and cumulatively similar geometric shapes of interacting orbitals.

II Systems *with organic isomorphic similarity* – systems which:

1) either differ by the number of heterogeneous atoms but have cumulatively similar geometric shapes of interacting orbitals;
2) or have certain differences in the geometric shape of orbitals but
   have the same number of interacting heterogeneous atoms.

III Systems without isomorphic similarity – systems considerably different both by the number of heterogeneous atoms and geometric shape of
their orbitals.

Taking into account the experimental data, all SEI types can be approximately classified as follows:

Systems I

1. $\alpha < (0-6)\%$; $\rho = 100\%$. Complete isomorphism, there is complete
   isomorphic replacement of atoms-components;
2. $6\% < \alpha < (25-30)\%$; $\rho = 98 - (0-3)\%$.

There is wide or unlimited isomorphism.

3. $\alpha > (25-30)\%$; no SEI.

Systems II

1. $\alpha < (0-6)\%$;
   a) There is reconstruction of chemical bonds that can be accompa-
      nied by the formation of a new compound;
   b) Cleavage of chemical bonds can be accompanied by a fragment
      separation from the initial structure but without adjoinings and
      replacements.
2. A $6\% < \alpha < (25-30)\%$; the limited internal reconstruction of chemical bonds is possible but without the formation of a new compound
   and replacements.
3. $\alpha > (20-30)\%$; no SEI.

Systems III

1. $\alpha < (0-6)\%$; a) The limited change in the type of chemical bonds
   in the given fragment is possible, there is an internal regrouping
   of atoms without the cleavage from the main molecule part and
   replacements;
   b) The change in some dimensional characteristics of the bond is
      possible;
2. A $6\% < \alpha < (25-30)\%$;

A very limited internal regrouping of atoms is possible;

   1. $\alpha > (25-30)\%$; no SEI.

When considering the above systems, it should be pointed out that they can be found in all cellular and tissue structures in some form but are not isolated and are found in spatial-time combinations.

The values of α and ρ calculated in such a way refer to a definite interaction type whose nomogram can be specified by fixed points of reference systems. If we take into account the universality of spatial-energy interactions in nature, such evaluation can have the significant meaning for the analysis of structural shifts in complex bio-physical and chemical processes of biological systems.

Fermentative systems contribute a lot to the correlation of structural interaction degree. In this model the ferment role comes to the fact that active parts of its structure (fragments, atoms, ions) have such a value of P_E-parameter, which equals the P_E-parameter of the reaction final product. That is the ferment is structurally "tuned" via SEI to obtain the reaction final product, but it will not enter it due to the imperfect isomorphism of its structure (in accordance with III).

The important characteristics of atom-structural interactions (mutual solubility of components, chemical bond energy, energy of free radicals, etc.) for many systems were evaluated following this technique [10, 11].

21.4 CALCULATION OF INITIAL DATA AND BOND ENERGIES

Based on the Eqs. (3)–(5) with the initial data calculated by quantum-mechanical methods [3–6] we calculate the values of P_0-parameters for the majority of elements being tabulated, constant values for each atom valence orbital. Mainly covalent radii – by the main type of the chemical bond of interaction considered were used as a dimensional characteristic for calculating P_E-parameter (Table 21.1). The value of Bohr radius and the value of atomic ("metal") radius were also used for hydrogen atom.

In some cases the bond repetition factor for carbon and oxygen atoms was taken into consideration [10]. For a number of elements the values of P_E-parameters were calculated using the ionic radii whose values are indicated in column 7. All the values of atomic, covalent and ionic radii were mainly taken by Belov-Bokiy, and crystalline ionic radii – by Batsanov [12].

The results of calculating structural P_S parameters of free radicals by the Eq. (15) are given in Table 21.2. The calculations are done for the radicals contained in protein and amino acid molecules (CH, CH_2, CH_3, NH_2 etc.), as well as for some free radicals formed in the process of radiolysis and dissociation of water molecules.

The technique previously tested [10] on 68 binary and more complex compounds was applied to calculate the energy of coupled bond of molecules by the equations:

$$\frac{1}{\mathring{A}} = \frac{1}{P_S} = \frac{1}{\left(P_E \dfrac{n}{K}\right)_1} + \frac{1}{\left(P_E \dfrac{n}{K}\right)_2} \; ; \qquad (16)$$

$$P_E \frac{n}{K} = P \qquad (17)$$

where n – bond average repetition factor, K – hybridization coefficient, which usually equals the number of registered atom valence electrons.

Here the P-parameter of energy characteristic of the given component structural interaction in the process of binary bond formation.

"Non-valence, nonchemical weak forces act ... inside biological molecules and between them apart from strong interactions" [13]. At the same time, the orientation, induction and dispersion interactions are used to be called Van der Waals. For three main biological atoms (nitrogen, phosphorus and oxygen) Van der Waals radii numerically equal approximately the corresponding ionic radii (Table 21.3).

It is known that one of the reasons of relative instability of phosphorus anhydrite bonds in ATP is the strong repulsion of negatively charged oxygen atoms. Therefore it is advisable to use the values of P-parameters calculated via Van der Waals radii as the energy characteristic of weak structural interactions of biomolecules (Table 21.3).

TABLE 21.1 P-parameters of atoms calculated via the bond energy of electrons.

Atom	Valence electrons	W (eV)	r_i (Å)	q^{2-}_0 (eVÅ)	P_0 (eVÅ)	R (Å)	P_0/R (eV)
						0.5292	9.0644
H	1S¹	13.595	0.5295	14.394	4.7985	0.28	17.137
						$R-_i$=1.36	3.525
	2P¹	11.792	0.596	35.395	5.8680	0.77	7.6208
						0.67	8.7582
	2P²	11.792	0.596	35.395	10.061	0.77	13.066
C						0.67	15.016
	2S²				14.524	0.77	18.862
	2S²+2P²				24.585	0.77	31.929
					24.585	0.67	36.694
	2P¹	15.445	0.4875	52.912	6.5916	0.70	9.4166
	2P²				11.723	0.70	16.747
	2P³	25.724	0.521	53.283	15.830	0.70	22.614
N						0.55	28.782
	2S²				17.833	0.70	25.476
	2S²+2P³				33.663	0.70	48.09
	2P¹	17.195	0.4135	71.383	6.4663	0.66	9.7979
	2P¹					R_i=1.36	4.755
	2P¹	17.195	0.4135	71.383	11.858	R_i=1.40	4.6188
	2P²					0.66	17.967
		17.195	0.4135	71.383	20.338	0.59	20.048
O	2P⁴					R_i=1.36	8.7191
		33.859	0.450	72.620	21.466	R_i=1.40	8.470
	2S²				41.804	0.66	30.815
	2S²+2P⁴					0.59	34.471
						0.66	32.524
						0.66	63.339
						0.59	70.854
	4S¹	5.3212	1.690	17.406	5.929	1.97	3.0096
	4S²				8.8456	1.97	4.4902
Ca	4S²					R^{2+}=1.00	8.8456
	4S²					R^{2+}=1.26	7.0203

TABLE 21.1 *(Continued)*

Atom	Valence electrons	W (eV)	r_i (Å)	q^2_0 (eVÅ)	P_0 (eVÅ)	R (Å)	P_0/R (eV)
	$3P^1$	10.659	0.9175	38.199	7.7864	1.10	7.0785
	$3P^1$					R^{3-}=1.86	P_3=4.1862
P	$3P^3$	10.659	0.9175	38.199	16.594	1.10	15.085
	$3P^3$					R^{3-}=1.86	8.9215
	$3S^2+3P^3$				35.644	1.10	32.403
Mg	$3S^1$	6.8859	1.279	17.501	5.8568	1.60	3.6618
	$3S^2$				8.7787	1.60	5.4867
						R^{2+}=1.02	8.6066
Mn	$4S^1$	6.7451	1.278	25.118	6.4180	1.30	4.9369
	$4S^1+3d^1$				12.924	1.30	9.9414
	$4S^2+3d^2$				22.774	1.30	17.518
Na		4.9552	1.713	10.058	4.6034	1.89	2.4357
	$3S^1$					R^{1+}_r=1.18	3.901
						R^{1+}_r=0.98	4.6973
K	$4S^1$	4.0130	2.612	10.993	4.8490	2.36	2.0547
						R^{1+}_r=1.45	3.344

TABLE 21.2 Structural P_S parameters calculated via the bond energy of electrons.

Radicals, molecule fragments	$P'_{ii}(eV)$	$P''_i(eV)$	$P_S(eV)$	Orbitals
OH	9.7979	9.0644	4.7080	O $(2P^1)$
	17.967	17.138	8.7712	O $(2P^2)$
H_2O	2×9.0644	17.967	9.0227	O $(2P^2)$
CH_2	17.160	2×9.0644	8.8156	C $(2S^1 2P^3_r)$
	31.929	2×17.138	16.528	C $(2S^2 2P^2)$
CH_3	15.016	3×9.0644	9.6740	C $(2P^2)$
	40.975	3×9.0644	16.345	C $(2S^2 2P^2)$
CH	31.929	12.792	9.1330	C $(2S^2 2P^2)$

TABLE 21.2   *(Continued)*

Radicals, molecule fragments	$P'_{ii}(eV)$	$P''_i(eV)$	$P_S(eV)$	Orbitals
NH	16.747	17.138	8.4687	N(2P^2)
	19.538	17.132	9.1281	N(2P^2)
NH$_2$	19.538	2×9.0644	9.4036	N(2P^2)
	28.782	2×17.132	18.450	N(2P^3)
CO–OH	8.4405	8.7710	4.3013	C(2P^2)
C=O	15.016	20.048	8.4405	C(2P^2)
C=O	31.929	34.471	16.576	O(2P^4)
CO=O	36.694	34.471	17.775	O(2P^4)
C–CH$_3$	17.435	19.694	9.2479	–
C–NH$_2$	17.435	18.450	8.8844	–
CO–OH	12.315	8.7712	5.1226	C(2S^{2}2P^2)
(HP)O$_3$	23.122	23.716	11.708	O(2P^2) P(3S^{2}3P^3)
(H$_3$P)O$_4$	17.185	17.244	8.6072	O(2P^1) P(3P^1)
(H$_3$P)O$_4$	31.847	31.612	15.865	O(2P^2) P(3S^{2}3P^3)
H$_2$O	2×4.3623	8.7191	4.3609	O(2P^2) r =1.36 Å
H$_2$O	2×4.3623	4.2350	2.8511	O(2P^2) r=1.40
C-H$_2$O	2.959	2.8511	1.4520	–
(C-H$_2$O)$_3$ Lactic acid	–	–	1.4520×3= 4.3563	–
(C-H$_2$O)$_6$ Glucose	–	–	1.4520×6= 8.7121	–

Bond energies for P and O atoms were calculated taking into account Van der Waals distances for atomic orbitals: $3P^1$ (phosphorus)-$2P^1$ (oxygen) and for $3P^3$ (phosphorus)-$2P^2$ (oxygen). The values of E obtained slightly exceeded the experimental, reference ones (Table 21.4). But for the actual energy physiological processes, for example, during photosynthesis, the efficiency is below the theoretical one, being about 83%, in some cases [14,15].

Perhaps the electrostatic component of resulting interactions at anion-anionic distances is considered in such a way. Actually the calculated value of 0.83E practically corresponds to the experimental values of bond energy during the phosphorylation and free energy of ATP in chloroplasts.

Table 21.4 contains the calculations of bond energy following the same technique but for stronger interactions at covalent distances of atoms for the free molecule P—O (sesquialteral bond) and for the molecule P=O (double bond).

The sesquialteral bond was evaluated by introducing the coefficient n = 1.5 with the average value of oxygen P_E parameter for single and double bonds.

The average breaking energy of the corresponding chemical bonds in ATP molecule obtained in the frameworks of semiempirical method PM3 with the help of software GAMESS [16] are given in column 11 of Table 21.4 for comparison. The calculation technique is detailed in Ref. [17].

The calculated values of bond energies in the system K—C-N being close to the values of high-energy bond P~O in ATP demonstrate that such structure can prevent the ATP synthesis.

When evaluating the possibility of hydrogen bond formation, we take into account such value of n/K in which $K=1$, and the value $n=3.525/17.037$ characterizes the change in the bond repetition factor when transiting from the covalent bond to the ionic one.

21.5 FORMATION OF STABLE BIOSTRUCTURES

At equilibrium-exchange spatial-energy interactions similar to isomorphism the electrically neutral components do not repulse but approach each other and form a new composition whose α in the Eqs. (13) and (14).

This is the first stage of stable system formation by the given interaction type which is carried out under the condition of approximate equality of component P-parameters: $P_1 \approx P_2$.

Hydrogen atom, element No 1 with the orbital $1S^1$ determines the main criteria of possible structural interactions. Four main values of its P-parameters can be taken from Tables 21.1 and 21.3:

1) for strong interactions: $P_E^{''} = 9.0644$ eV with the orbital radius 0.5292 Å and $P_E^{'''} = 17.137$ eV with the covalent radius 0.28 Å.

2) for weaker interactions: $P_E^{'} = 4.3623$ eV and $P_E = 3.6352$ eV with Van der Waals radii 1.10 Å and 1.32 Å. The values of P-parameters $P' : P'' : P'''$ relates as 1:2:4. In accordance with the concepts in Section 2, such values of interaction P-parameters define the normative functional states of biosystems, and the intermediary can produce pathologic formations by their values.

The series with approximately similar values of P-parameters of atoms or radicals can be extracted from the large pool of possible combinations of structural interactions (Table 21.5). The deviations from the initial, primary values of P-parameters of hydrogen atom are in the range ± 7%.

The values of P-parameters of atoms and radicals given in the Table define their approximate equality in the directions of interatomic bonds in polypeptide, polymeric and other multiatom biological systems.

In ATP molecule these are phosphorus, oxygen and carbon atoms, polypeptide chains – CO, NH and CH radicals. In Table 21.5 you can also see the additional calculation of their bond energy taking into account the sesquialteral bond repetition factor in radicals C—O и N—H.

On the example of phosphorus acids it can be demonstrated that this approach is not in contradiction with the method of valence bonds, which explains the formation peculiarities of ordinary chemical compounds. It is demonstrated in Table 21.6 that this electrostatic equilibrium between the oppositely charged components of these acids can correspond to the structural interaction for H_3PO_4—$3P^1$ orbitals of phosphorus and $2P^1$ of oxygen, and for HPO_3—$3S^23P^3$ orbitals of phosphorus and $2P^2$ of oxygen. Here it is stated that P-parameters for phosphorus and hydrogen subsystems are added algebraically. It is also known that the ionized phosphate groups are transferred in the process of ATP formation that is apparently defined for phosphorus atoms by the transition from valence-active $3P^1$ orbitals to $3S^2 3P^3$ ones, that is, 4 additional electrons will become valence-active. According to the experimental data the synthesis of one ATP molecule is connected with the transition of four protons and when the fourth proton is being transited the energy accumulated by the ferment reaches its threshold [18, 19]. It can be assumed that such proton transitions in ferments initiate similar changes in valence-active states in the system P–O. In the process of oxidating phosphorylation the transporting ATP-synthase uses the energy of gradient potential due to $2H^+$-protons, which, in the given model for such a process, corresponds to the initiation of valence-active transitions of phosphorus atoms from $3P^1$ to $3P^3$-state.

In accordance with the Eq. (17) we can assume that in stable molecular structures the condition of the equality of corresponding effective interaction energies of the components by the couple bond line is fulfilled by the following equations:

$$\left(P_E \frac{n}{K} \right)_1 \approx \left(P_E \frac{n}{K} \right)_2 \rightarrow P_1 \approx P_2 \tag{18}$$

And for heterogeneous atoms (when $n_1 = n_2$):

$$\left(\frac{P_E}{K} \right)_1 \approx \left(\frac{P_E}{K} \right)_2 \tag{18a}$$

In phosphate groups of ATP molecule the bond main line comprises phosphorus and oxygen molecules. The effective energies of these atoms by the bond line calculated by the equation (18) are given in Tables 21.4 and 21.5, from which it is seen that the best equality of P_1 and P_2 parameters is fulfilled for the interactions $P(3P^3) - 8.7337$ eV and $O(2P^2) - 8.470$ eV that is defined by the transition from the covalent bond to Van der Waals ones in these structures.

The resulting bond energy of the system P–O for such valence orbitals and the weakest interactions (maximum values of coefficient K) is 0.781 eV (Table 21.4). Similar calculations for the interactions $P(2P^1) - 4.0981$ eV and $O(2P^1) - 4.6188$ eV produce the resulting bond energy 0.397 eV.

The difference in these values of bond energies is defined by different functional states of phosphorous acids HPO_3 and H_3PO_4 in glycolysis processes and equals 0.384 eV that is close to the phosphorylation value (0.34–0.35 eV) obtained experimentally.

Such ATP synthesis is carried out in anaerobic conditions and is based on the transfer of phosphate residues onto ATP via the metabolite. For example: ATP formation from creatine phosphate is accompanied by the transition of its NH group at ADP to NH_2 group of creatine at ATP.

TABLE 21.3   Ionic and Van der Waals radii (Å).

Atom	Ionic radii			Van der Waals radii		
	Orbital	R_I	P_E/κ (eV)	R_B	Orbital	P_E/κ (eV)
H	1S¹	R=1.36	3.525	1.10	1S¹	4.3623
		r =0.5292	9.0644	1.32		3.6352
N	2P³	R³=1.48	10.696/3=3.5653	1.50	2P¹	4.3944/1
				1.50	2P³	10.553/3=3.5178
				1.50	2S²2P³	22.442/5=4.4884
P	3P³	R³= 1.86	8.9215/3=2.9738	1.9	3P¹	4.0981/1
				1.9	3P³	8.7337/3=2.9112
				1.9	3S²3P³	18.760/5=3.752
O	2P²	R²=1.40	8.470/2=4.2350	1.40	2P¹	4.6188/1
				1.50	2P¹	4.3109/1
		R²=1.36	8.7191/2=4.3596	1.40	2P²	8.470/2=4.2350
				1.50	2P²	7.9053/2=3.9527
C	2S²2P²	d*/2=3.2/2=1.6	15.365/4=3.841	1.7	2P¹	3.4518/1
				1.7	2P²	5.9182/2=2.9591
				1.7	2S²2P²	14.462/4=3.6154

d* – contact distance between C–C atoms in polypeptide chains [13].

TABLE 21.4 Bond energy (eV).

Atoms, structures, orbitals	Bond	Remarks		Component 2		Component 3		Calculation			Remarks
		P_E (eV)	n/K	P_E (eV)	n/K	P_E (eV)	n/K	E	E [13, 14, 15]	E [16][17]	
1	2	3	4	5	6	7	8	9	10	11	12
P–O $3S^23P^3-2S^22P^4$	cov.	32.403	1.5/5	70.854 / 63.339	1.5/6 / 1.5/6	6.14		6.277 / 5.024 / <6.15>	6.1385 / 6.14		PO free molecule
H_2O $1S^12P^2$	cov. / cov.	2×9.0624 / 2×9.0624	1/1 / 1/1	17.967 / 20.048	1/6 / 2/2			2.570 / 9.520	2.476	10.04	Decay of one molecule
H_3PO_4	cov	3×9.0624	1/1	32.405	1/5	4×17.967	1/2	4.8779	4.708		
C–O $(2P^11S^1)$	cov.	7.6208	1.125/2	9.7979	1/1			4.2867			
C–N $2P^1-2P^1$	cov.	7.6208	1/4	9.4166	1/5			0.9471			
C–N $2P^1-2P^1$	cov.	7.6208	1.125/4	9.4166	1.1667/5			1.0898	0.870		
K-C-N $4S^22P^1-2P^1$	cov.	2.0547	1/1	7.6208	1/4	9.4166	1/5	0.648			
$(C-H_2O)-(C-H_2O)$	VdW	1.4520	1/1	1.4520	1/1			0.726			

TABLE 21.4 (Continued)

Atoms, structures, orbitals	Bond	Remarks P_E (eV)	n/K	Component 2 P_E (eV)	n/K	Component 3 P_E (eV)	n/K	Calculation E	E [13, 14, 15]	E [16] [17]	Remarks
C–O $2S^2 2P^2$-$2P^2$	cov.	31.929 31.929	1.125/4 1/4	20.048 20.042	1/2 1/2			4.7367 4.4437			
N–H $2P^1$-$1S^1$	cov.	9.4166 9.4166	1.1667/1 1/1	9.0644 9.0644	1/1 1/1			4.9654 4.6186			
C–H $2P^1$-$1S^1$	cov.	13.066	1/2	9.0644	1/1			3.797	3.772		
C–H $2P^2$-$1S^1$		13.066	1/2	17.137	1/1			4.7295			
N–H$_2$ $2P^3$-$1S^1$	cov.	22.614	1/3	2 × 9.0644	1/1			5.3238			
–H···O		3.525	3.525/17.037	4.6188	1/6			0.3730	0.3742		Hydrogen bond
P–O $3P^3$-$2P^2$	cov.	15.085	2/3	20.042	2/2			6.6970	6.504	6.1385	Free molecule
P–O $3P^3$-$2P^2$	VdW	8.7337	1/5	8.470	1/6			0.781	0.670		ΔG ATP
P–O $3P^1$-$2P^1$	cov.	7.0785	1/1	9.7979	1/1			4.1096	4.2059	4.2931	
P–O $3P^1$-$2P^1$	VdW	4.0981	1/5	4.6188	1/6			0.3970	0.34–0.35		Phospholyration

TABLE 21.5 Bio-structural spatial-energy parameters (eV).

Series No	H	C	N	O	P	CH	CO	NH	Glucose	Lactic acid	OH	Remarks
I	9.0644 ($1S^1$)	8.7582 ($2P^1$) 9.780 ($2P^1$)	9.4166 ($2P^1$)	9.7979 ($2P^1$)	8.7335 ($3P^1$)	9.1330 ($2S^22P^11S^1$)	8.4405 ($2P^2$-$2P^2$)	8.4687 ($2P^2$-$1S^1$) 9.1281 ($2P^2$-$1S^1$)	8.7121 $2P^-$ $-(1S^1$-$2P^2)$		8.7710	Strong interaction
II	17.132 ($1S^1$)	17.435 ($2S^12P^1$)	16.747 ($2P^2$)	17.967 ($2P^2$)	18.760 ($3S^23P^2$)	C and H blocks	16.576 ($2S^22P^2$- $2P^1$)	N and H blocks				Strong interaction
III	(4.3623) ($1S^1$)	3.8696 ($2P^2$)	4.3944 ($2P^1$)	4.3109 ($2P^1$) 4.6188 ($2P^1$)	4.0981 ($3P^1$)	4.7295	4.4437 4.7367	4.6186 4.9654		4.3563 $2P^-$ ($1S^1$- $2P^2$)	4.7084	Weak interaction
IV	3.6352 ($1S^1$)	3.4518 ($2P^1$) 3.6154 ($2S^22P^2$)	3.5178 ($2P^1$)	4.2350 ($2P^2$) 3.6318 ($2P^1$)	4.0981 ($3P^1$) 3.752 ($3S^23P^1$)	4.7295	4.4437 4.7367	4.6186 4.9654				Effective bond energy

TABLE 21.6   Structural interactions in phosphorus acids.

Molecule	Component 1			Component 2			$\alpha = (DP/<P>)$ *100%
	Atom	Orbitals	$P=P_1+P_2$ (eV)	Atom	Orbitals	P(eV)	
$(H_3P)O_4$	H_3P	$1S^{1-}3P^1$	$4.3623*3+ 4.0981$ $=17.185$	O_4	$2P^1$	$4.3109*4$ $=17.244$	0.34
		$1S^{1-}$		at	$2P^2$		0.74
		$(3S^23P^3)$	$4.3623*3+18.760$ $=31.847$	$r=1.50$Å		$7.9053*4$ $=31.612$	
$(HP)O_3$	HP	$1S^1 -$ $(3S^23P^3)$	$4.3623+18.760$ $=23.122$	O_3 at $r=1.50$Å	$2P^2$	$7.9053*3$ $=23.716$	2.54

From Table 21.4, it is seen that the change in the bond energy of these two main radicals of metabolite is 5.3238–4.9654 = 0.3584 eV – taking the sesquialteral bond N⋯H into account (as in polypeptides) and 5.3238–4.6186 = 0.7052 eV – for the single bond N-H. This is one of the intermediary results of the high-energy bond transformation process in ATP through the metabolite. From Tables 21.4 and 21.6 we can conclude that the phosphorous acid H_3PO_4 can have two stationary valence-active states during the interactions in the system P–O for the orbitals with the values of P-parameters of weak and strong interactions, respectively. This defines the possibility for the glycolysis process to flow in two stages. At the first stage, the glucose and H_3PO_4 molecules approach each other due to similar values of their P-parameters of strong interactions (Table 21.2). At the second stage, H_3PO_4 P-parameter in weak interactions 4.8779 eV (Table 21.4) in the presence of ferments provokes the bond $(H_2O–C)–(C–H_2O)$ breakage in the glucose molecule with the formation of two molecules of lactic acid whose P-parameters are equal by 4.3563 eV. The energy of this bond breakage process equaled to 0.726 eV (Table 21.4) is realized as the energy of high-energy bond it ATP.

According to the reference data about 40% of the glycolysis total energy, that is, about 0.83 eV, remains in ATP.

By the hydrolysis reaction in ATP in the presence of ferments ($HPO_3 + H_2O \rightarrow H_3PO_4 + E$) for structural P_S parameters (Table 21.2) E = 11.708 + 4.3609–15.865 = 0.276 eV.

It is known that the change in the free energy (ΔG) of hydrolysis of phosphorous anhydrite bond of ATP at pH = 7 under standard conditions is 0.311–0.363 eV. But in the cell the ΔG value can be much higher as the ATP and ADP concentration in it is lower than under standard conditions. Besides, the ΔG value is influenced by the concentration of magnesium ions, which is the acting conferment in the complex with ATP. Actually Mg^{2+} ion has the P_E parameter equaled to 8.6066 eV (Table 21.1), which is very similar to the corresponding values of P-parameters of phosphorous and oxygen atoms.

The quantitative evaluation of this factor requires additional calculations.

21.6   CONCLUSIONS

1. Bond energies of some biostructures have been calculated following P-parameter and quantum-mechanical techniques.
2. High-energy bonds in ATP are formed in the system P—O under functional transitions of their valence-active states.
3. The data obtained agree with the experimental ones.

KEYWORDS

- **experimental data**
- **functional transitions**
- **high-energy bonds in ATP**
- **phosphorus-oxygen**
- **quantum-mechanical data**
- **spatial-energy concept**

REFERENCES

1. Eyring, H., Walter, J., Kimball, G. E. Quantum chemistry. I. L., M., 1998, 528p.
2. Korablev, G. A. Spatial-Energy Principles of Complex Structures Formation. Brill Academic Publishers and VSP, Netherlands, 2005, 426pp. (Monograph).
3. Fischer, C. F. Average-Energy of Configuration Hartree-Fock Results for the Atoms Helium to Radon. Atomic Data, 1972, № 4, pp. 301–399.
4. Waber, J. T., Cromer, D. T. Orbital Radii of Atoms and Ions. J. Chem. Phys., 1965, V 42, №12, pp. 4116–4123.
5. Clementi, E., Raimondi, D. L. Atomic Screening constants from, S. C. F. Functions, 1. J. Chem. Phys., 1963, V. 38, №11, pp. 2686–2689.
6. Clementi, E., Raimondi, D. L. Atomic Screening constants from, S. C. F. Functions, 1. J. Chem. Phys., 1997, V. 47, №14, pp. 1300–1307.
7. Gombash, P. Atom statistical model and its application. M.: I. L., 1951, 398 p.
8. Clementi, E. Tables of atomic functions. J. B. M. S. Re. / Develop. Suppl., 1995, V. 9, №2, 76.
9. Korablev, G. A., Zaikov, G. E. Spatial-Energy Parameter as a Materialised Analog of Wafe Function. Progress on Chemistry and Biochemistry, Nova Science Publishers, Inc. New York, 2009, 355–376.
10. Korablev, G. A., Zaikov, G. E. Energy of chemical bond and spatial-energy principles of hybridization of atom orbitalls. J. of Applied Polymer Science, V. 101, №3, Ang. 5, 2006, pp. 2101–2107.
11. Korablev, G. A., Zaikov, G. E. Formation of carbon nanostructures and spatial-energy criterion of stabilization. Mechanics of composite materials and structures, 2009, RAS, v. 15, №1, 106–118.
12. Batsanov, S. S. Structural chemistry. Facts and dependencies. M.: MSU, 2009.
13. Volkenshtein. Biophysics. M.: Nauka, 1988, 598 p.
14. Photosynthesis / Ed. by Govindzhi. M.: Mir, v. 1, 1987, 728 p; v. 2, 1987, 460 p.
15. Clayton, R. Photosynthesis. Physical mechanisms and chemical models. M.: Mir, 1984, 350 p.
16. Schmidt, M. W., Baldridge, K. K., Boatz, J. A. et al. General atomic and molecular electronic structure system. J. Comput. Chem., 1993, v. 14, p. 1347–1363.
17. Khokhriakov, N. V., Kodolov, V. I. Influence of active nanoparticles on the structure of polar liquids. Chemical physics and mesoscopy, 2009, v. 11. № 3, p. 388–402.
18. Feniouk, B. A. Study of the conjugation mechanism of ATP synthesis and ATP proton transport. Referun – Biology and natural science, 1998, 108 p.
19. Feniouk, B. A., Junge, W., Mulkidjanian, A. Tracking of proton flow across the active ATP-synthase of Rhodobacter capsulatus in response to a series of light flashes. EBEC Reports, Volume 10, p. 112, 1998.

CHAPTER 22

ULTRASONIC NANOMEDICINE IN THE ASPECT OF CANCER THERAPY

A. L. NIKOLAEV, A. V. GOPIN, V. E. BOZHEVOLNOV,
N. V. ANDRONOVA, E. M. TRESCHALINA, and
N. V. DEZHKUNOV

CONTENTS

ABSTRACT

In this chapter, the author's views on the use of solid inclusions in biological structures as "concentrators" of acoustic energy for ultrasound treatment of cancer (solid-phase sonosensitization) are presented. Particular attention is drawn to the possibility of synthesis of these inclusions directly in the tumor. The validity of the hypothesis of solid-phase sonosensitization is confirmed in experiments on model systems and animals.

22.1 INTRODUCTION

Works in which nanoparticles (micelles, liposomes, bubbles, nanocapsules), introduced into the bloodstream, are the means of delivering drugs to the tumor, and ultrasound is a factor stimulating drug release [1–3] and enhancing the therapeutic effect [4, 5], can be attributed to the field of ultrasonic nanomedicine.

The method of ultrasonic nanotherapy of malignant tumors, developed in this study, differs from those described in the literature by the statement, that nanoparticles and their aggregates are formed immediately in the tumor from nontoxic and nonmedicinal precursors. It is provided by biochemical features of its growth. As a result of the metabolic atypia, physicochemical conditions in the tumor (decreased pH, an increased content of calcium ions in intercellular liquid, monotonicity of the lipidic structure of membranes) differ from conditions in normal tissues surrounding the tumor. These differences result in the possibility of solid phase formation mostly in the tumor. The solid phase segregates in the tumor after intravenous introduction of solutions of compounds whose calcium salts or acidic forms are insoluble under tumor conditions. Thus, the selectivity of the formation of nanoparticles and their aggregates mostly in the tumor can be achieved using the least specific, hence, most stable symptoms of its atypia.

The ultrasound-induced therapeutic effect on the biological systems modified by nanoparticle aggregates is achieved due additional acoustic energy release in regions where these aggregates are localized. This occurs due to the fact that aggregates locally change the ultrasound absorbance, enhancing thermal effects and increasing the intensity of cavitation processes. If these aggregates are localized in tissue or blood vessels of the tumor, then additional acoustic energy release in these areas can cause the death of the tumor or slow down its growth.

The above considerations were put into the basis of the development of the method of ultrasonic tumor destruction in the presence of solid nanoparticles and their aggregates. We called the phenomena underlying this method and associated with the presence of the solid phase as the solid-phase sonosensitization, and nanoparticles themselves and their aggregates as solid-phase sonosensitizers (SPSs) by analogy with known sonosensitizers [6].

Over the years experiments on model gel systems imitating tumor tissue [7], bacterial cells and experimental animals [8–10] were carried out to assess the practical significance of enhancement of ultrasound effects through the introduction of SPSs and optimization of the conditions of the solid-phase sonosensitization method.

In this chapter, we studied the effect of solid-phase modifiers on the thermal and cavitation effects occurring in polymer hydrogels on ultrasound exposure. Evaluation of thermal effects was carried out thermometric. Activity of cavitation processes was assessed by measuring the level of scattered noise, and on information about destruction of polymer matrix of the hydrogel. The effect of solid-phase sonosensitization was tested in experiments in vitro on bacterial cells and in vivo on mice.

22.2 EXPERIMENTAL

22.2.1 SOLID-PHASE MODIFIERS

Theraphathal™ – octasodium salt of cobalt octacarboxyphthalocyanine. Theraphthal is soluble in water compound. It was synthesized in Organic Intermediates and Dyes Institute (Russia). Calcium salt of Theraphthal (insoluble in water) was precipitated from water solution by slow adding of calcium chloride solution to Theraphthal solution while stirring.

Silica gel and hydrophobized by long-chain (C_{16}) carboxylic acids silica gel were produced by BioChemMack (Russia).

22.2.2   GELS

A pluronic gel was prepared by dissolving solid Pluronic F127 (Sigma-Aldrich, Germany) in cooled (4°C) water in concentration 20 wt. %. If necessary, the solid-phase modifier was added (silica gel or hydrophobized silica gel) in the required concentration. The resulting solution undergoes a transition at a critical point when heated to 30°C, accompanied by gelation.

An agarose gel was prepared by dissolving solid agarose (Difco, United States) in water on heating to 90°C in a concentration of 1.5 wt %. If necessary, the solid-phase modifier was added (calcium salt of Theraphthal) in the required concentration. Then, the solution was slowly cooled to 20°C, which resulted in its gelation.

Agarose hydrogels were also modified by successive impregnation, first, with Theraphthal solution (0.01 mol/L) and then with calcium chloride solution (1 mol/L). This resulted in precipitation of solid phase of calcium salt of Theraphthal within the sample. Impregnation with each of the solutions lasted for 48 h. Then gel samples were placed in a large amount of water for 48 h to remove any remaining reagent. Similarly, precipitation of iron hydroxide (FeOOH) was carried out. Sample was impregnated, first, with a 0.1 mol/L solution of iron (III) chloride (Reachim, Russia) and, then, with a 4 mol/L aqueous ammonia (Reachim, Russia). The estimated concentration of precipitated solid phase was 1 wt. % in both cases.

22.2.3   THERMAL EFFECTS

To estimate the thermal effects of an ultrasound field, an ultrasound emitter and a cylindrical sample (2.5 cm high and 3 cm in diameter) were positioned coaxially and immersed into a thermostated vessel with degassed water. A thermocouple was inserted from one sample end to the center along the symmetry axis. Ultrasound was fed from another end using a planar piezoceramic emitter 2 cm in diameter with resonant radiation at a frequency of 2.64 MHz with an intensity of 2 W/cm^2 at the emitter. The sample was placed at a distance of 20 cm from the emitter.

22.2.4   CAVITATIONAL EFFECTS OF ULTRASOUND FIELD

In these studies, we used unfocused ultrasound with a frequency of 0.88 MHz and an intensity of 2 W/cm^2. Cavitation properties of gels containing impurities and without them were comparatively estimated using an IS-3 MS cavitation indicator developed at the Belarusian State University of Informatics and Radioelectronics. The device's operating principle is based on the measurement of the level of scattered noise in a frequency band from 2 to 10 MHz, received by a broadband hydrophone. An emitter 2 cm in diameter was in direct contact with the face of the cylindrical gel sample (2 cm high and 2.5 cm in diameter), that is, measurements were performed in the transducer's near field. The hydrophone with a sensitive element 2 mm in diameter was placed in the immediate vicinity from the side surface of the sample.

The cavitational destruction of Pluronic F127 molecules after ultrasonic treatment (0.88 MHz 1 W/cm^2) was estimated by the molecular weight distribution measured by gel permeation chromatography. Treated gel samples were cooled in ice and then diluted with water. Solid-phase modifiers were separated by centrifugation. The resulting solutions were freeze-dried, and molecular weight distributions were determined for extracted polymers.

22.2.5   IN VITRO EXPERIMENTS

2 mL of the suspension of bacterial cells and 2 mL of Theraphthal solution (10^{-5} mol/L) were introduced into the thermostated vessel with an ultrasonic transparent bottom and were exposed to ultrasonic treatment (0.88 MHz, 1 W/cm^2) for 10 min at 38°C.

22.2.6 ELECTRON MICROSCOPY

Small samples of hydrogels (1 mm^3) or bacteria cells on membrane filters were places in liquid propane for 30 s and then in cooled to $-95°C$ acetone, which was gradually heated to room temperature. The resulting material was dried in critical point dryer (Hitachi HCP-2). Then samples were coated with gold in an ion coater (EIKO IB-3) and were studied under a JEOL JEM-100B microscope at an accelerating voltage of 15 kV.

22.2.7 IN VIVO EXPERIMENTS

Melanoma B16 was inoculated intramuscularly into BDF1 mice in the right paw according to the standard procedure [8, 10]. The initial tumor volume at 8th day after inoculation was $V_0 = 1.1 \pm 0.1$ cm^3. Sonosensitizers were injected intravenously 1 h prior to ultrasonic treatment. Ultrasonic treatment of inoculated tumors was performed simultaneously with two frequencies (0.88 MHz 1 W/cm^2 and 2.64 MHz 2 W/cm^2) for 10 min at 40°C. The dynamics of tumor growth was assessed by the change in its volume.

22.3 RESULTS

22.3.1 MODIFICATION OF HYDROGELS

Phases of insoluble compounds (calcium salt of Theraphthal and iron hydroxide) were precipitated in an agarose hydrogel (Figs. 22.1 and 22.2).

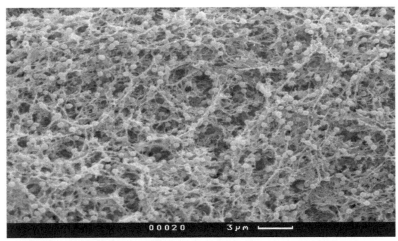

FIGURE. 22.1   Electron micrograph of agarose gel modified with calcium salt of Theraphthal, synthesized directly in the gel (1 wt. %).

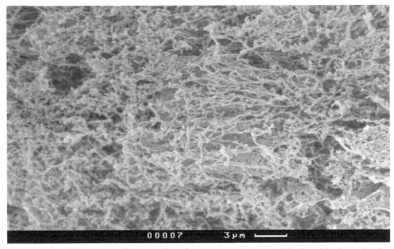

FIGURE 22.2   Electron micrograph of agarose gel modified with iron hydroxide, synthesized directly in the gel (1 wt. %).

These data indicate the presence of two types of solid phase localization in the gel. In one case, the crystals are located at individual centers of polymer matrix (calcium salt of Theraphthal, (Fig. 22.1), in another, they are uniformly distributed over the matrix (iron hydroxide, (Fig. 22.2). This affects the magnitude of the thermal effects of ultrasound exposure.

22.3.2   THERMAL EFFECTS

Thermal effects of ultrasound have been studied on samples of agarose gel modified with various modifiers. Figure 22.3 shows the dynamics of the temperature growth of agarose gel modified with calcium salt of Theraphthal on ultrasound exposure.

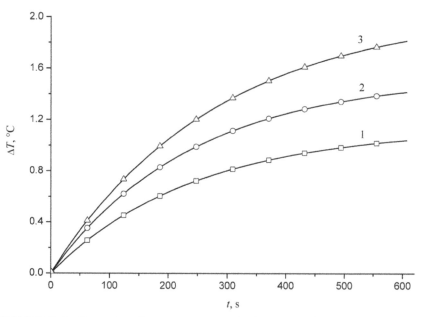

FIGURE 22.3   Dynamics of temperature growth of agarose gel modified with calcium salt of Theraphthal in an ultrasonic field: (1) gel without a modifier, (2) modifier introduced by mechanical mixing while gelation of the gel, (3) modifier synthesized within the gel.

Data show that the introduction of the calcium salt of Theraphthal increases the thermal effects. It should be noted that the magnitude of the thermal effect depends on the method of introduction of a modifier into gel.

Under the same conditions there is no increase of the thermal effect for the iron hydroxide. In some cases there is even a slight decrease in the absorption of ultrasound. This can be caused by a change in mechanical characteristics of the matrix network in these samples, which is inlaid with the highly dispersed solid phase of iron hydroxide and changes its viscoelastic characteristics.

22.3.3 CAVITATIONAL EFFECTS

Agarose gel without modifier and gel containing 1 wt. % of calcium salt of Theraphthal, precipitated in gel, were used in the experiments. The experimental results are shown in Fig. 22.4, where the vertical axis is the noise level recorded by the IS-3 MS indicator, the horizontal axis is the ultrasound intensity. We can see that the level of the signal measured at an intensity of 2 W/cm² in the modified hydrogel sample (U_1) significantly (by a factor of more than 4) exceeds the corresponding value in gel without modifier (U_2). This indicates that the conditions for occurrence and development of cavitation in modified gels significantly more favorable. Also, a possible increase in the concentration of gas bubbles in modified gels can also enhance the thermal effects of ultrasound exposure.

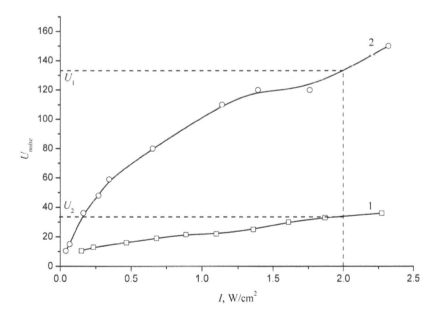

FIGURE 22.4 Comparative estimate of the broadband noise level measured in a frequency band of 2–10 MHz under exposure agarose gel samples to an ultrasonic frequency of 0.88 MHz, depending on the ultrasound intensity: (1) unmodified gel and (2) gel modified with calcium salt of Theraphthal.

Effect of SPSs on the cavitational destruction of polymer molecules (Pluronic F127) was estimated by changes in the molecular weight distribution and polydispersity in modified and unmodified gel samples after ultrasonic treatment (0.88 MHz, 1 W/cm^2). The results are presented in Table 22.1.

TABLE 22.1   Effect of ultrasound treatment on the molecular weight distribution of Pluronic F127.

Sample	M_n, kDa	M_w, kDa	M_w/M_n
Pluronic gel	10.3±0.1	13.4±0.1	1.30±0.02
Pluronic gel after ultrasound treatment	9.4±0.1	12.0±0.1	1.28±0.02
Pluronic gel with silica gel after ultrasound treatment	8.6±0.1	11.6±0.1	1.35±0.02
Pluronic gel with hydrophobized silica gel after ultrasound treatment	7.1±0.1	10.8±0.1	1.52±0.02

In the presence of SPSs molecular weight distribution is shifted to lower weights (decrease in number average M_n and weight average M_w molecular weight) and polydispersity is increased (increase in the ratio M_w/M_n). These facts are signs of polymer molecules destruction. At the same time, the molecular weight distribution of the unmodified polymer after ultrasonic treatment is practically unchanged. These results indicate the increasing of the intensity of the ultrasonic degradation processes in the presence of SPSs.

22.3.4   EXPERIMENTS IN VITRO

It was shown on model biological systems (*Enterococcus spp.* bacteria) that the destructive effect of ultrasound is enhanced in the presence of nanoparticles. Figure 5 shows scanning electron micrographs of bacteria after the combined action of ultrasound and Theraphthal.

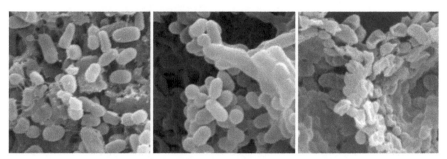

FIGURE 22.5   Electron micrographs of *Enterococcus spp.* cells: untreated cells (left), after ultrasonic treatment (center) and after combined action of ultrasound and Theraphthal (right).

It is evident that as a result of ultrasonic treatment some of the bacteria exposed to destruction apparently accompanied by the leakage of cytoplasm. However, most of the cells keep normal form. Bacteria pretreated with Theraphthal show a change in the shape and the destruction of the membranes that cover almost all the treated cells. This is confirmed by data on the survival of bacteria. Enhancing of effect of ultrasound treatment of cells is associated with the formation of calcium salts of Theraphthal solid phase on the membranes, thereby reducing the mechanical strength of membranes and enhancing the local cavitational effects.

22.3.5   EXPERIMENTS IN VIVO

Preclinical trials of the method of solid-phase sonosensitization were carried out at N.N. Blokhin Russian Research Oncological Center. The experiments included estimation of the therapeutic efficiency, harmlessness, and the effect on metastatic disease. These studies showed a high therapeutic efficiency of the method, that is, tumor regression by 75–80% on average with an increase in the animal lifetime by a factor to 2, good exposure tolerance, and the absence of the effect on metastatic disease. Currently, the method of ultrasonic therapy with solid-phase sonosonosensitizers is in clinical trials.

Figure 22.6 shows the tumor growth dynamics in several experimental series using Theraphthal, octasodium salt of zinc octacarboxyphthalocyanine (ZnPc), and gold nanoparticles stabilized with polyethylene glycol.

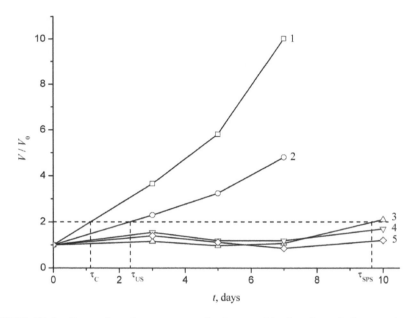

FIGURE 22.6 Dynamics of tumors growth after combined action of ultrasound and SPSs. V_0 is the tumor volume to the beginning of ultrasound exposure, V is the volume of the tumor after a time T after exposure. (1) control; (2) ultrasound; (3) Theraphthal, 30 mg/kg + ultrasound; (4) gold nanoparticles, 7 mg/kg + ultrasound; (5) ZnPc, 12 mg/kg + ultrasound.

The particles of SPSs were deposited directly in the tumor after intravenous injection of Theraphthal or ZnPc solutions. Gold nanoparticles were injected intravenously in the form of suspension. It is evident that the tumor doubling time in the experiments using SPSs (τ_{SPS}) increases ten times in comparison with the control group (τ_C) and five times in comparison with the case of ultrasound exposure alone (τ_{US}).

22.4 CONCLUSION

Analysis of the results shows that the solid-phase inclusion in polymeric and biological structures are effective local "amplifiers" of thermal and cavitational effects of ultrasound exposure. This can be used in the therapy of cancer. Selection of the optimal method of administration of SPSs in the tumor is determined by the specific conditions: the type of tumor, its localization, the relative toxicity of SPSs, duration of procedure, etc. In our opinion, the method of synthesis of sensitizers directly in the tumor may be in some cases easier to implement in the clinical and less toxic.

KEYWORDS

- cavitation
- nanoparticles
- solid sonosensitizers
- sonosensitization
- therapy of cancer
- ultrasound
- ultrasound therapy

REFERENCES

1. Gao, Z.G., Fain, H.D., Rapoport, N. Controlled and targeted tumor chemotherapy by micellar-encapsulated drug and ultrasound, Journal of Controlled Release, 2005, 102(1), 203–222.
2. Rapoport, N. Physical stimuli-responsive polymeric micelles for anticancer drug delivery, Progress in Polymer Science, 2007, 32(8–9), 962–990.
3. Chumakova, O.V., Liopo, A.V.,reev, V.G., Cicenaite, I., Evers, B.M., Chakrabarty, S., Pappas, T.C, and Esenaliev, R.O. Composition of PLGA and PEI/DNA nanoparticles improves ultrasound-mediated gene delivery in solid tumors in vivo, Cancer Letters, 2008, 261(2), 215–225.
4. Harada Yo., Ogawa, K., Irie Yu., Endo, H., Feril Jr. L.B., Uemura, T., Tachibana, K. Ultrasound activation of TiO2 in melanoma tumors, Journal of Controlled Release, 2011, 149, 190–195.
5. Yamaguchi Sh., Kobayashi, H., Narita, T., Kanehira, K., Sonezaki Sh., Kudo, N., Kubota Yo., Terasaka Sh., Houkin, K. Sonodynamic therapy using water-dispersed TiO_2 polyethylene glycol compound on glioma cells: Comparison of cytotoxic mechanism with photodynamic therapy, Ultrasonics Sonochemistry, 2011, 18, 1197–1204.
6. Rosenthal, I., Sostaric, J.Z., Riesz, P. Sonodynamic therapya review of the synergistic effects of drugs and ultrasound, Ultrasonics Sonochemistry, 2004, 11(6), 349–363.

7. Nikolaev, A.L., Gopin, A.V., Chicherin, D.S., Bozhevol'nov, V.E., Melikhov, I.V. Lo-calization of Acoustic Energy in Gel Systems on Solid-Phase Inhomogeneities, Moscow University Chemistry Bulletin, 2008, 63(3), 167–171.
8. Andronova, N.V., Filonenko, D.V., Bozhevolnov, V.E., Nikolaev, A.L., Treshalin, I.M., Treshalina, H.M., Gerasimova, G.K., Kaliya, O.L., Vorozhtsov, G.N. Anticancer therapy combined with local ultrasound (experimental evaluation), Russian Journal of Biotherapy, 2005, 4(3), 101–105 (in Russian).
9. Nikolaev, A.L., Gopin, A.V., Bozhevolnov, V.E.,ronova, N.V., Treschalina, E.M., Gerasimova, G.K., Khorosheva, E.V., Kaliya, O.L., Vorozhtsov, G.N., Dezhkunov, N.V. Sonodynamic cancer therapy: integrated experimental study, Materials of III Euro-asian congress on medical physics, Russia Moscow 21–26 June 2010, 1, 150–152 (in Russian).
10. Andronova, N.V., Bokhyan, B.Yu., Nikolaev, A.L., Treschalina, E.M., Aliev, M.D., Kovalevskiy, E.E., Filonenko, D.V., Gopin, A.V., Bozhevolnov, V.E., Kogan, B.Ya., Vorozhtsov, G.N. Rationale for clinical study of preoperative treatment of soft tissue sarcomas with local ultrasound hyperthermia and chemotherapy with cisplatin and/or doxorubicin, Sarcoma of bone, soft tissue and skin tumors, 2011, 1, 28–33 (in Russian).

CHAPTER 23

A DETAILED REVIEW ON PRODUCTION OF ELECTROSPUN CNT-POLYMER COMPOSITE NANOFIBERS

M. HASANZADEH, V. MOTTAGHITALAB, R. ANSARI, B. HADAVI MOGHADAM, and A. K. HAGHI

CONTENTS

ABSTRACT

Polymer nanofibers are being increasingly used for a wide range of applications owing to their high specific surface area. Electrospinning process, as a novel and effective method for producing nanofibers from various materials, has been used to fabricate nanofibrous membrane. Carbon nanotubes (CNTs) have a number of outstanding mechanical, electrical, and thermal properties, which make them attractive as reinforcement in polymer matrix. Incorporation of chapter provides a comprehensive review of current researches and developments in the field of electrospun CNT-polymer composite nanofiber with emphasis on the processing, properties, and application of composite nanofiber as well as the theoretical approaches on predicting mechanical behavior of CNT-polymer composites. The current limitations, research challenges, and future trends in modeling and simulation of electrospun polymer composite nanofibers are also discussed.

23.1 INTRODUCTION

With the rapid development and growing role of nanoscience and nanotechnology in recent years, considerable research efforts have been directed towards the development and characterization of fibrous materials with diameter in the range of tens to hundreds of nanometers [1–4].

Electrospinning, as a simple and powerful technique for producing ultrafine fibers, has evinced more interest and attention in recent years [5–10]. In this process, the nanofibers are generated by the application of a strong electric field on a polymer solution or melt. The submicron-range spun fibers produced by this technique possess high specific surface area, high porosity, and small pore size [11–15]. Electrospun polymer nanofibers are very attractive multifunctional nanostructures due to its versatility and potential for diverse applications. Some of these notable applications include tissue engineering scaffolds [16–18], biomedical agents [19], protective clothing [20], drug delivery [21], super capacitors [22, 23], and energy storage [24].

Carbon nanotubes (CNTs) are highly desirable materials possessing unique structural, mechanical, thermal, and electrical properties [25–30]. Electrospun nanotube-polymer composite nanofibers are very attractive materials for a wide range of applications. This is due to the fact that the use of the electrospinning technique to incorporate CNTs in polymer nanofibers induces alignment of nanotubes within the nanofiber structure, which could greatly enhance the mechanical, electrical and thermal properties of composite fibers.[31–34]

Numerous studies have focused on understanding and improving the structure and properties of the electrospun CNT-polymer composites [35–45]. However, due to the difficulties encountered in experimental characterization of nanomaterial, the simulation and theoretical approaches play a significant role in understanding the properties and mechanical behavior of CNT-reinforced polymer nanofibers [46].

The current review summarizes the recent progress made in electrospun CNT-polymer composite nanofiber, along with their processing, characterization, mechanical properties, and applications. Theoretical investigations on mechanical properties of CNT and CNT-based polymer composite are also addressed. Finally, research challenges and future trends in modeling and simulation of electrospun polymer composite nanofibers are discussed.

23.2 FUNDAMENTAL OF ELECTROSPINNING

23.2.1 CONCEPTS AND MECHANISM

Electrospinning, as a straightforward, simple and effective method for preparation of nanofibrous materials, have attracted increasing attention during the last two decade [5–10]. Electrospinning process, unlike the conventional fiber spinning systems (melt spinning, wet spinning, etc.), uses electric field force instead of mechanical force to draw and stretch a polymer jet [47]. This process involves three main components including syringe filled with a polymer solution, a high voltage supplier to provide the required electric force for stretching the liquid jet, and a grounded collection plate to hold the nanofiber mat. A schematic representation of electrospinning setup is shown in Fig. 23.1. In electrospinning process, when the electric field overcomes the surface tension force of the droplet of the polymer solution formed on the tip of the syringe, the charged polymer solution forms a liquid jet and travels towards collection plate. The ejected polymer solution forms a conical shape known as the "Taylor cone" and is drawn towards a grounded collection plate [48–60].

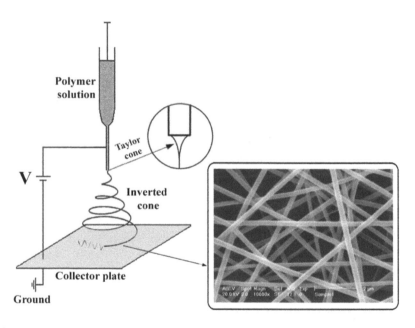

FIGURE 23.1   Schematic drawing of the electrospinning process.

Electrospinning process can be explained by four major regions including the Taylor cone region, the stable jet region, the instability region, and the base region. Once the electric field reaches a critical value, a charged jet of the solution is ejected from the tip of the Taylor cone and will begin to thin due to the forces acting on it. The thinning of the jet can be divided into two different stages. The initial stage is a period of thinning as a straight jet and the later stage is a period of thinning due to the bending/whipping instability. Typically, electrospinning process has four types of physical instability: the classical Rayleigh instability, the axisymmetric instability, the bending instability, which results in whipping, and the whipping instability. These instabilities influence the morphology and structure of the deposited fibers [61–65]. Figure 23.2 illustrates the four types of instability of the jet.

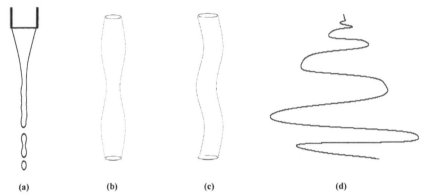

(a) (b) (c) (d)

FIGURE 23.2 Instabilities in the jet (a) Rayleigh instability (b) axisymmetric conducting instability (c) bending instability (d) whipping instability.

Rayleigh instability is axisymmetric and is commonly observed in low electric field strength (low charge densities) or when the viscosity of the solution is below the optimum value. This instability causes liquid jet to break-up due to surface tension force. At high electric fields, the Rayleigh instability disappears and is replaced by a bending instability. Both the bending and whipping (nonaxisymmetric) instabilities occur owing to the charge-charge repulsion between the excess charges present in the jet. The bending instability produces oscillations in the diameter of the jet in the axial direction caused by electrical forces (see Fig. 23.2). The nonaxisymmetric instability replaces the axisymmetric (i.e., Rayleigh and bending) instabilities at higher electric fields and produces bending and stretching force on the jet [61].

The morphology and the structure of the electrospun nanofibers are dependent upon many parameters, which are mainly divided into three categories: solution properties, processing parameters, and ambient conditions, as illustrated in Fig. 23.3 [66–68].

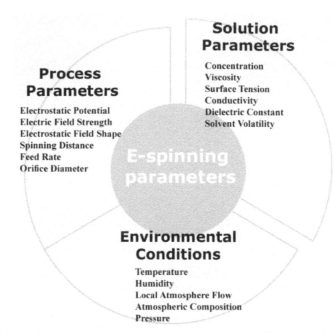

FIGURE 23.3   Electrospinning parameters that are known to affect the resultant nanofiber morphology.

23.2.2   GOVERNING EQUATIONS

The analysis of electrospinning process is based on the slender-body theory. It is widely used in fiber spinning of viscoelastic liquid. To simplify the mathematical description, a few idealizing assumptions are made. The jet radius R decreases slowly along the axial direction $Z : |dR(Z) / dZ| \ll 1$. Furthermore it is assumed that the fluid velocity υ is uniform in the cross section of the jet.

The basic governing equations for the stable jet region are the equation of continuity, conservation of electric charges, linear momentum balance, and electric field equation. As main source for these flow equations we refer to Refs. [1, 62, 69].

The most important and simplest relation is the equation of continuity; it describes the conservation of mass in electrospinning

$$\pi R^2 v = Q \tag{1}$$

where Q is a constant volume flow rate. The conservation of electric charge may be expressed by

$$\pi R^2 KE + 2\pi R v\sigma = I \tag{2}$$

where k is the electrical conductivity of the liquid jet, E is the axial component of the electric field, σ is the surface charge density, and I is the constant total current in the jet. The conservation of momentum for the fluid is formulated by (see Fig. 23.4).

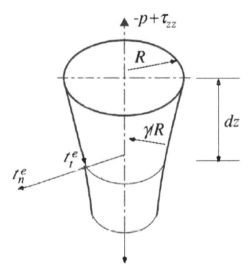

FIGURE 23.4 Momentum balance on a short section of the jet. Adapted from Ref. [62].

$$\frac{d}{dz}(\pi R^2 \rho v^2) = \pi R^2 \rho g + \frac{d}{dz}[\pi R^2(-p + \tau_{zz})] + \frac{\gamma}{R}.2\pi RR' + 2\pi R(t_t^e - t_n^e R') \tag{3}$$

where ρ is the fluid density, g is the acceleration due to gravity, p is the pressure, τ_{zz} is the axial viscous normal stress, γ is the surface tension, and t_t^e and t_n^e are the tangential and normal tractions, respectively, on the surface of the jet due to electricity. The prime indicates a derivative with respect to z, and R' is the slope of the jet surface. The ambient pressure has been set to zero. The electrostatic tractions are determined by the surface charge density and the electric field:

$$t_n^e = \left\| \frac{\varepsilon}{2}(E_n^2 - E_t^2) \right\| \approx \frac{\sigma^2}{2\bar{\varepsilon}} - \frac{\bar{\varepsilon} - \varepsilon}{2}E^2 \tag{4}$$

$$t_t^e = \sigma E_t \approx \sigma E \tag{5}$$

where ε and $\bar{a}$ are the dielectric constants of the jet and the ambient air, respectively, E_n and E_t are the normal and tangential components of the electric field at the surface, and $\|*\|$ indicates the jump of a quantity across the surface of the jet. We have used the jump conditions for E_n and E_t: $\|\mathring{a}E_n\| = \bar{a}\bar{E}_n - \varepsilon E_n = \sigma$, $\|E_t\| = \bar{E}_t - E_t = 0$, and assumed that $\mathring{a}E_n \ll \bar{a}\bar{E}_n$ and $E_t \approx E$. The overbar indicates quantities in the surrounding air. The pressure $p(z)$ is determined by the radial momentum balance, and applying the normal force balance at the jet surface leads to

$$-p + \tau_{rr} = t_n^e - \frac{\gamma}{R} \tag{6}$$

Inserting Eqs. (4)–(6) into Eq. (3) yields

$$\rho v v' = \rho g + \frac{3}{R^2}\frac{d}{dz}(\eta R^2 v') + \frac{\gamma R'}{R^2} + \frac{\sigma \sigma'}{\varepsilon} + (\varepsilon - \bar{\varepsilon})EE' + \frac{2\sigma E}{R} \tag{7}$$

where η is the fluid viscosity and may depend on the local strain rate or the accumulated strain. The equation of electric field requires that

$$E = E_\infty - \left[\frac{1}{\tilde{a}} (\sigma R)' - \left(\frac{\varepsilon}{\tilde{a}} - 1 \right) \frac{(ER^2)''}{2} \right] \ln \left(\frac{d}{R_0} \right) \tag{8}$$

where E_∞ is the applied external electric field, d is the distance between the nozzle and the collector, and R_0 is the radius of the spinneret.

23.3 CARBON NANOTUBE (CNT)

23.3.1 CNT STRUCTURE

Carbon nanotubes are classified as single-walled nanotubes (SWNTs) and multiwalled nanotubes (MWNTs). A SWNT can be visualized as a grapheme sheet that has been rolled into a hollow cylinder with ends caps. So the atomic structure of SWNT can be described by the chiral vector C_h and the chiral angle θ (see Fig. 23.5) given by Refs. [70–77]:

$$\vec{C}_h = n\vec{a_1} + m\vec{a_2} \tag{9}$$

where a_1 and a_2 are unit vectors and integers (n, m) are the number of steps along the zigzag carbon bonds. The tubes with $n=m$ are known as armchair tubes and $m=0$ as zigzag tubes. In all other combinations of n and m, the SWNT are known as chiral tubes. Figure 23.6 illustrates the schematic of nanotubes with different chiralities, including armchair and zigzag nanotubes. The chiral angle θ is the angle made by the chiral vector C_h with respect to the zigzag direction $(n, 0)$ and is defined as:

$$\theta = \cos^{-1} \frac{(2n+m)}{2\sqrt{(m^2 + mn + n^2)}} \tag{10}$$

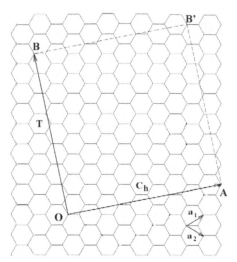

FIGURE 23.5   Schematic illustration of a hexagonal grapheme sheet and definitions of SWNTs parameters.

The radius of nanotube R is given by following equation:

$$R = \frac{a\sqrt{(m^2 + mn + n^2)}}{2\pi} \tag{11}$$

where a is a unit vector length and $a = a_{C-C}\sqrt{3}$, in which a_{C-C} is the carbon–carbon bond length and is equal to 0.1421 nm. Considering the effective wall thickness of SWNT t, the effective radius R_{eff} is defined as:

$$R_{eff} = \frac{a\sqrt{(m^2 + mn + n^2)}}{2\pi} + \frac{t}{2} \tag{12}$$

To relate the grapheme atomic coordinates and SWNT from a planer hexagonal lattice, the following equation is used:

$$(X, Y, Z) = [R\cos(\frac{x}{R}), R\sin(\frac{x}{R}), y] \tag{13}$$

In which X, Y and Z represent the nanotube coordinates and x and y are the grapheme coordinates. Table 23.1 summarizes the characteristic parameters of carbon nanotube.

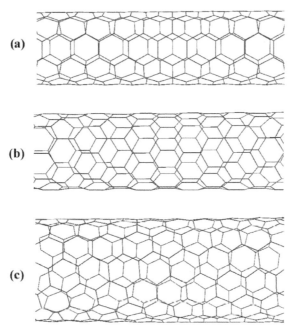

FIG. 23.6 Schematic representation of (a) armchair (5,5), (b) zigzag (9,0), and (c) chiral (8,6) nanotubes.

TABLE 23.1 Characteristic parameters of carbon nanotube.

Symbol	Description	Formula	Value
a_{C-C}	Carbon-carbon bond length	—	0.1421 nm
a	Unit vector length	$a = a_{C-C}\sqrt{3}$	0.2461 nm
a_1, a_2	Unit vectors	$\left(\dfrac{\sqrt{3}}{2}, \dfrac{1}{2}\right)a, \left(\dfrac{\sqrt{3}}{2}, -\dfrac{1}{2}\right)a$	In (x, y) coordinate system
C_h	Chiral vector	$C_h = n\,a_1 + m\,a_2$	(n, m): integer
L	Circumference of nanotube	$L = \left|C_h\right| = a\sqrt{m^2 + mn + n^2}$	

TABLE 23.1   *(Continued)*

Symbol	Description	Formula	Value
R	Radius of nanotube	$R = \dfrac{a\sqrt{(m^2 + mn + n^2)}}{2\pi}$	
R_{eff}	Effective radius of nanotube	$R_{eff} = \dfrac{a\sqrt{(m^2 + mn + n^2)}}{2\pi} + \dfrac{t}{2}$	
θ	Chiral angle	$\theta = \cos^{-1}\dfrac{(2n + m)}{2\sqrt{(m^2 + mn + n^2)}}$ $\theta = \sin^{-1}\dfrac{\sqrt{3}m}{2\sqrt{(m^2 + mn + n^2)}}$ $\theta = \tan^{-1}\dfrac{\sqrt{3}m}{2n + m}$	$0° \leq \theta \leq 30°$

23.3.2 CNT PROPERTIES

It is well known that CNTs possess remarkable properties and have a combination of outstanding mechanical, electrical, thermal properties, and low density, so they have been suggested as excellent candidate for numerous applications. Some of these physical properties are summarized in Table 23.2 [78–80]. It is claimed that CNTs are a hundred times stronger than steel and have very high Young's modulus, which are the most important parameters that define the mechanical stiffness of a material. The actual mechanical properties measurements of CNTs were made on individual arc discharge MWNTs using atomic force microscope (AFM). The Young's modulus values of 270–970 GPa was obtained for a range of MWNTs [81]. Measurements on SWNT showed a tensile modulus of ~1000 GPa for small diameter SWNT bundles by bending methods. However, the properties of larger diameter bundles were dominated by shear slippage of individual nanotubes within the bundle [25].

TABLE 23.2 Physical properties of CNTs.

Physical property	Unit	CNT	
		SWNT	MWNT
Aspect ratio	—	100–10000	100–10,000
Specific gravity	(g/cm^3)	0.8	1.8
Specific surface area	(m^2/g)	10–20	—
Elastic modulus	(GPa)	~1000	300–1000
Tensile strength	(GPa)	5–500	10–60
Thermal stability in air	(°C)	>600	>600
Thermal conductivity	(W/mK)	3000–6000	2000
Electrical conductivity	(S/cm)	$10^2–10^6$	$10^3–10^5$
Electrical resistivity	$(\mu\Omega/cm)$	5–50	—

In addition to their outstanding mechanical properties, CNTs exhibit exceptionally high thermal and electrical conductivity. However, several parameters affect these properties, including the synthesis methods employed, defects, chirality, degree of graphitization and diameter [82]. For example, depending on chirality, the CNT can be metallic or semiconducting. The remarkable thermal conductivity of CNTs makes them particularly attractive for thermal management in composites. It is known that thermal properties of CNTs play critical roles in controlling the performance of CNT based composite materials. Similarly, CNTs possess high electrical properties, which suggest their use in miniaturized electronic component [83–85].

23.4 ELECTROSPUN CNT-POLYMER COMPOSITE

Incorporation of CNT into polymer matrix, due to the exceptional properties and large aspect ratio, has been proven to be a promising approach leading to structural materials and composites with excellent physical and mechanical properties such as tensile strength, tensile modulus, strain to failure, torsional modulus, compressive strength, glass transition temperature, solvent resistance, and reduced shrinkage [86]. Fibrous materials were found to be most suitable for many applications. There are various techniques for the fabrication of CNT-polymer composites, such as solution casting, melt processing, melt spinning, electrospinning, and in-situ polymerization [31]. Electrospinning as an effective processing method to produce CNT-polymer nanofibers with the CNTs orienting to the axes of the as-spun nanofibers have attracted increasing attention during the last two decades.

23.4.1   BASIC PRINCIPLES

As the alignment of CNTs in the polymer matrix is an interesting field, and plays an important role on unidirectional properties such as strength, modulus, and toughness, the electrospinning has been widely used to make CNTs align along the fiber axis. Due to the improved CNT alignment within the nanofibers and simple spinning process, this technique has been regarded as the most promising approach for producing electrospun carbon nanotube-polymer composites [69, 87]. In this technique, initially the CNTs are randomly oriented, but they are aligned with the flow of polymer (see Fig. 23.7). The CNTs alignment has been analyzed based on the planar sink flow in a wedge, also known as Hamel flow [88]. This model implies that gradual alignment of random CNTs into the fiber occurs at the central streamline of the Taylor cone. While nanotube center moves along the streamline, they were drawn towards the tip of the wedge [89].

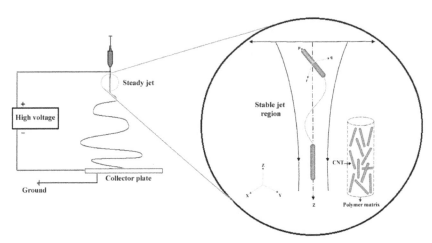

FIGURE 23.7 Schematic illustration of CNT alignment in jet flow.

A number of reports on the electrospun composite fibers using various types of polymers and CNTs are listed in Table 23.3. Although CNTs have potential to be embedded into various polymer matrices, electrospinning of some polymers causes difficulties.

TABLE.23.3 Electrospun CNT-polymer composite in the literature.

Type of CNTs	Polymers	Molecular weight (M_w)	Solvents	CNTs (wt.%)	Focus of the research	Ref.
SWNT	PVA	146,000–186,000, (96% hydrolyzed)	—	10	Structure-property relationships	[37]
MWNT	PVA	89,000–98,000, (99+% hydrolyzed)	Water/ethanol (3:1)	4	Effect of carbon nanotube aspect ratio and loading on elastic modulus	[90]
MWNT	PVA	75,000–79,000 (98–99% hydrolyzed)	Water	20 (v/v)%	Morphology and mechanical properties	[91]
MWNT	PTh	—	Chloroform	—	Electric and dielectric properties	[92]

TABLE 23.3 *(Continued)*

Type of CNTs	Polymers	Molecular weight (M_w)	Solvents	CNTs (wt.%)	Focus of the research	Ref.
MWNT	PAN	—	DMF	0.25	Effect of conductive additive and filler on the process	[93]
MWNT	PAN	—	DMF	0–10	Anisotropic electrical conductivity	[94]
MWNT	PAN	70,000	DMF	2	Application of vibration technology in electrospinning	[95]
MWNT	PAN	150,000	DMF	1	Monitoring the interaction between the π-electrons of CNT and the nitrile groups of PAN by using synchrotron microbeam WAXD analysis	[96]
MWNT	PAN	—	DMF	1	Creation of continuous yarn and characterization of the surface morphology, and the mechanical properties	[97]
SWNT	PAN/ MA/ itaconic acid (93:5.3:1.7w/w)	100,000	DMF	0–1	Thermal and tensile properties	[98]

TABLE 23.3 *(Continued)*

Type of CNTs	Polymers	Molecular weight (M_w)	Solvents	CNTs (wt.%)	Focus of the research	Ref.
MWNT	PAN/ MA/ itaconic acid (93:5.3:1.7w/w)	100,000	DMF	0–3	Investigating the distribution and alignment of CNT	[99]
MWNT	PMMA	350,000	DMF	5	Tensile mechanics	[100]
SWNT	PMMA	996,000	chloroform	0–1	Temperature dependent electrical resistance and morphology	[101]
MWNT	Poly(ε-caprolactone)	~100,000	DMF	3	Effects of green tea polyphenols (GTP) and MWNTs incorporation on nanofiber morphology, mechanics, in vitro degradation and in vitro GTP release behaviors.	[102]
MWNT	PS	185,000	DMF/ THF (2:3)	0.8, 1.6	Production of hollow nanofibers with un-collapsing and surface-porous structure	[103]
SWNT, DWNT, MWNT	PS	—	DMF	0–5	Morphology, structure and properties	[42]
MWNT	PLA	180,000	DMF	1–5	the electrical conductivity, mechanical properties and in vitro degradation stability	[104]

TABLE 23.3 *(Continued)*

Type of CNTs	Polymers	Molecular weight (M_w)	Solvents	CNTs (wt.%)	Focus of the research	Ref.
SWNT	PLA/PAN	—	DMF	—	Fabrication of continues CNT-filled nanofiber yarn	[35]
MWNT	PLLA/PCL	100,000	Chloroform/ methanol (≈3:1)	0–3.75	Morphology, mechanical properties, in vitro degradation and biocompatibility	[105]
MWNT	PCL	—	Chloroform/ methanol (3:1)	2–3	Morphology and structural properties	[106]
MWNT	PC	—	—	15	Characterizing the composite nanofibers with directly embedded CNTs, with respect to the orientation and uniform dispersion of CNTs within the electrospun fibers	[107]
CNT	Alginate	—	Water	0–1	Mechanical and Electrical Properties	[43]
MWNT	PET	19,200	TFA	0–3	Tensile, thermal, and electrical properties	[108]

TABLE 23.3 *(Continued)*

Type of CNTs	Polymers	Molecular weight (M_w)	Solvents	CNTs (wt.%)	Focus of the research	Ref.
MWNT	PA 6,6	—	FA/DCM (2:1)	0–2.5	Effect of fiber diameter on the deformation behavior of self-assembled electrospun CNT-PA 6,6 fibers	[109]
MWNT	PVDF/PPy	—	DMAc	1	Morphology, chemical structure, electrical conductivity, mechanical and thermal properties	[110]
MWNT	PEO	900,000	Ethanol/water (40:60)	50 (v/v)%	Morphology and mechanical properties	[91]
MWNT	PANi/PEO	PANi=65,000 PEO= 600,000	Chloroform	—	Electro-magnetic interference shielding	[111]
MWNT	PANi/PEO	PANi=65,000 PEO= 100,000	Chloroform	0.25–1	Electrical conductivity	[112]
MWNT	PPy	—	—	15	The electro-chemistry and current-voltage characteristics	[113]
MWNT	PBT	—	HFIP	5	Morphology and mechanical properties	[114]

Acronyms: DWNT: Double-walled nanotube; PVA: Poly(vinyl alcohol); PTh: Poly(thiophene); PAN: Poly(acrylonitrile); MA: Methyl acrylate; PMMA: Poly(methyl methacrylate); PS: Polystyrene; PLA: Poly DL-lactide; PC: Polycarbonate; PCL:

Polycaprolactone; PET: Poly(ethylene terephthalate); PA: Polyamide; PVDF: Polyvinylidene fluoride; PPy: Polypyrrole; P(VDF-TrFE): Poly(inylidene difluoride-trifluoroethylene); PANi: Polyaniline; PEO: Poly(ethylene oxide); PU: Polyurethane; PBT: Poly(butylene terephthalate); DMF: *N, N*-dimethylformamide; THF: Tetrahydrofuran; TFA: Trifluoro acetic acid; FA: Formic acid; DCM: Dichloromethane; DMAc: *N, N*-dimethylacetamide; HFIP: Hexafluoro-2-propanol.

* The molecular weight of the polymer is an estimate from literature. Also note where not specified, the average molecular weight is M_w.

23.4.2   *STRUCTURAL AND MORPHOLOGICAL PROPERTIES*

It is found that most semicrystalline polymers could be crystallized during the fiber formation process. During the electrospinning process, a fraction of the polymer chains crystallizes into lamellae or small crystallites, and another fraction remains amorphous [5]. The relaxed amorphous tie molecules exist between the crystalline parts of the chain. Figure 23.8 represents the expected general structure in CNT-polymer composite fiber. Due to the shear and elongation forces acting on the jet, the tie molecules pass through the neighboring crystallites to form small-sized bundles and the lamellae are rearranged to form fibrils [115]. The crystalline and amorphous fractions of the chains within the electrospun fibers influence the physical and mechanical properties of the nanofibers. The amorphous phase corresponds to the elastomeric properties and the crystalline phase provides dimensional stability. It seems that the CNTs embedded in electrospun fibers reduce the overall mobility of the polymer chains. Thus, orientation of polymer chains during electrospinning and the presence of CNTs within the fibers enhance the structural properties of the electrospun CNT-polymer composite [86].

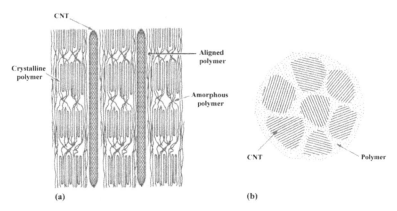

FIGURE 23.8 (a) Longitudinal and (b) Cross-sectional representation of CNT polymer composite fiber.

It is known that rapid solvent evaporation from the electrospinning jet accompanied by the rapid structure formation and leads to a decrease in the jet temperature. Therefore, the aligned molecules have less time to realign themselves along the fiber axis, leading to less developed structure in the fiber.

There are two methods to study the morphological properties of electrospun CNT-polymer composite: one is electron microscopy (EM) and the second one is atomic force microscopy (AFM). As mentioned earlier, the final morphologies of the electrospun CNT-polymer composite fibers can be affected by several characteristics of the initial solution such as solution concentration, CNT weight fraction, viscosity, surface tension and conductivity of solution in addition to some electrospinning process (applied voltage, spinning distance, volume flow rate, and the strength of the applied electric field) and environmental conditions (temperature and humidity).

23.5 MECHANICAL PROPERTIES OF ELECTROSPUN CNT-POLYMER COMPOSITE

23.5.1 BACKGROUND

The exceptional structure of CNTs, their low density, their high aspect ratio, and outstanding mechanical and physical properties make them an

ideal candidate for specific applications in which CNTs are used as reinforcements in composite materials. There are several parameters affect the
mechanical properties of composite materials, including large aspect ratio
of reinforcement, good dispersion, alignment and interfacial stress transfer
[80, 82]. These factors are discussed below.

The CNT aspect ratio is one of the main effective parameters on the
longitudinal elastic modulus. Generally, the CNTs have a high aspect ratio,
but their ultimate performance in a polymer composite is different. Stress
transfer from the polymer matrix to the dispersed CNTs increases with increase in the aspect ratio of CNT [116]. However, the aggregation of CNTs
could lead to a decrease in the effective aspect ratio of the CNTs.

The uniformity and stability of nanotube dispersion in polymer matrix
are probably the most fundamental issue for the performance of composite materials. A good dispersion and distribution of CNTs in the polymer
matrix minimizes the stress concentration centers and improves the uniformity of stress distribution in composites [80]. On the other hand, if the
nanotubes are poorly dispersed within the polymer matrix, the composite
will fail because of the separation of the nanotube bundle rather than the
failure of the nanotube itself, resulting in significantly reduced strength
[117]. Mazinani et al. studied the CNT dispersion for electrospun composite fiber, as well as its effect on the morphologies and properties of electrospun CNT-polystyrene nanocomposite [42]. They demonstrated that
the CNT dispersion is an important controlling parameter for final fibers
diameter and morphology.

Another parameter influencing the mechanical properties of nanotube
composites is the CNT alignment. The effects of CNT alignment on electrical conductivity and mechanical properties of CNT-polymer nanocomposites have been discussed in a number of researches [118–120]. For
example, it has been reported that with increasing CNT alignment, the
electrical and mechanical properties of the SWNT-epoxy composites increased due to an increased interface bonding of CNTs in the polymer
matrix [119].

A good interfacial adhesion between the matrix and the nanotubes is another effective parameter of CNT-polymer composites. There are some possible adhesions between CNT and polymer matrix, including physical, chemical and/or mechanical. It is known that diffusive and electrostatic adhesions

are not common in polymer composites. One of the most common types of adhesion in polymer composites is physical adhesion, such as van der Waals force, that refers to the intermolecular forces between CNT and polymer matrix. Chemical adhesion as the strongest form of adhesion represents chemical bonding between the CNT and polymer matrix. The interlocking and entanglement of CNT functional chains and polymer matrix can be represented by mechanical adhesion. It should be mentioned that nanotubes and nanofibers due to their perfect cylinders with smooth surface exhibit insignificant mechanical interlocking with polymer matrix [80] (Fig. 23.9).

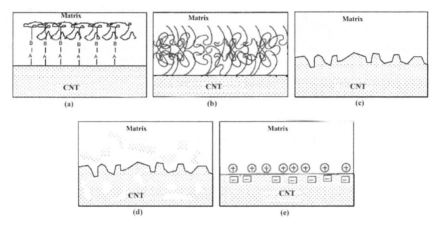

FIGURE 23.9 Schematic representation of (a) chemical bonding, (b) molecular entanglement, (c) mechanical interlocking, (d) diffusion adhesion, and (e) electrostatic adhesion between the matrix and the CNTs.

Other parameters influencing the mechanical properties of nanotube composites include: solvent selection, size, crystallinity, crystalline orientation, purity, entanglement, and straightness [82].

23.5.2   CNT-POLYMER INTERACTIONS

In CNT-polymer composites, the constituents retain their own original chemical and physical identities but together they produce a combination of properties that cannot be achieved by any of the components acting alone [121]. Properties of composite materials greatly depend on the nature of bonding at the interface, the mechanical load transfer from the surrounding matrix to the nanotube and the strength of the interface. The interface plays an important role in optimizing load transfer between the CNT and the polymer matrix. The mechanism of interfacial load transfer from the matrix to nanotubes can be categorized as: micromechanical interlocking, chemical bonding, and the weak van der Waals force between the matrix and the reinforcement [122, 123]. As reported previously, if the load stress cannot be effectively transferred from the matrix to the CNTs, the physical and mechanical properties of the nanocomposites could be considerably lower than the expected. In order to improve the mechanical properties of nanocomposite materials, strong interfacial interaction between the nanotubes and the polymer matrix is a necessary condition, but might not be a sufficient condition.

In addition to interfacial interactions between the CNT and the polymer matrix, the dispersion of CNTs in the polymer has significant influence on the performance of a CNT-polymer nanocomposite. Many different approaches have been used by researchers in an attempt to disperse CNT in polymer matrix such as physical sonication and chemical modification of CNT surface [124–126]. Functionalization of CNT surface can lead to the construction of chemical bonds between the nanotube and polymer matrix and offers the most efficient solution for the formation of strong interface. A strong interface between the coupled CNT-polymer creates an efficient stress transfer [31]. It should be noted that covalent functionalization of CNT may disrupt the grapheme sheet bonding, and thereby reduce the mechanical properties of the final product. However, noncovalent treatment of CNT can improve the CNT-polymer (Fig. 23.10) composite properties through various specific interactions [127].

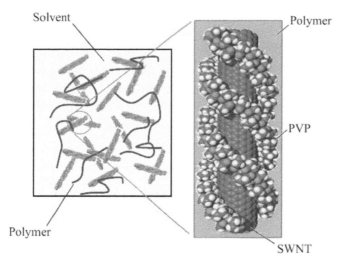

FIGURE 23.10   Schematic representation of functionalized CNT.

23.5.3   MEASUREMENTS

According to the literature, most studies on CNT-polymer composites have been reported the enhancement of mechanical strength by CNT addition, and this is of particular importance for electrospun nanofibers. It is well known that measuring the mechanical properties of individual electrospun nanofibers, due to the small size, is difficult. In this regard, a few experimental investigations reported their findings on measuring mechanical properties of single electrospun composite nanofiber [128–132].

Several nanomechanical characterization techniques have been suggested to measure the elastic properties of individual electrospun nanofibers, such as AFM cantilevers, universal tensile tester, and AFM-based nanoindentation system [133, 134]. Among them AFM-based techniques have been widely used to measure the mechanical properties of single electrospun nanofibers. This technique was carried out by attaching nanofiber to two AFM cantilever tips and recording the cantilever resonances for both the free cantilever vibrations and for the case where the microcantilever system has nanofibers attached. The Young's modulus of the nanofiber is then derived from the measured resonant frequency shift resulting from the nanofiber. A similar experiment has been carried out for electrospun polyethylene oxide (PEO) nanofibers [130] by using a piezo-resistive AFM tip. In this chapter, the selected nanofiber was first attached to a piezo-resistive AFM cantilever tip and to a movable optical microscope stage. The nanofiber was stretched by moving the microscope stage and the force applied to the nanofiber was measured via the deflection of the cantilever. These techniques are suitable for fibers with diameter ranging from several micrometers to tens of nanometers. However, it is difficult to manipulate and test individual nanofibers.[133]

23.6   ELECTROSPUN CNT-POLYMER COMPOSITE APPLICATIONS

The research and development of electrospun nanofibers has evinced more interest and attention in recent years due to the heightened awareness of its potential applications in the medical, engineering and defense fields [135–140]. Despite the several published reviews on polymer nanofibers applications, a few investigations have been reported on applications of electrospun CNT-polymer composites. However, most of works on electrospun CNT-polymer composite fibers have focused on developing a fundamental understanding of the composite structure-property relationships. Electrospun CNT-polymer composite nanofibers, due to their excellent mechanical, thermal, and electrical properties, as well as nanometer scale diameter, are appropriate for a large variety of potential applications, such as in military protective clothing, in fuel cells, in nanosensors and in energy storage. Figure 23.11 shows the potential applications of electrospun composite nanofibers [140–146].

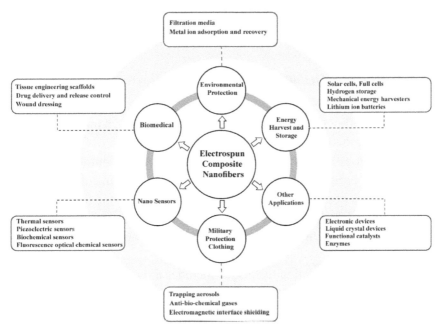

FIGURE 23.11   Potential applications of electrospun composite nanofibers.

In a recent investigation, electrospun chitosan/PVA nanofibers rein-forced by SWNT (SWNT-CS/PVA) have been used as scaffold for neural tissue engineering [147]. The in vitro biocompatibility of the electrospun fiber mats was also assessed using human brain-derived cells and U373 cell lines. The obtained results indicated that SWNTs as reinforcing phase can augment the morphology, porosity, and structural properties of CS/PVA nanofiber composites and thus benefit the proliferation rate of both cell types. Moreover, the cells exhibit their normal morphology while integrating with surrounding fibers [147]. In another study, electrospun MWNT-PANi/PEO nanofibers were found to enhance the electromagnetic interface (EMI) shielding properties of nanofibers [148]. It is found that absorption was the main reaction that shielded the electromagnetic inter-ference. Electrospun CNTs-polymer nanofibers have also shown potential applications in filter media, tissue engineering scaffolds, wound dressing, face masks, drug delivery, enzyme carriers, super capacitors, and compos-ite reinforcement [135, 149–151].

23.7 THEORETICAL APPROACHES IN CNT-POLYMER COMPOSITE

23.7.1. BACKGROUND

Due to the difficulties encountered in experimental characterization of nanomaterial, theoretical approaches in determining the CNT-based composite properties have attracted a great interest. A tremendous amount of research has been developed to understand the mechanical behavior of CNTs and their composites. According to the literature, the theoretical and computational approaches can be mainly classified into two categories: i) atomistic modeling, and ii) continuum mechanics approaches [152–164] (see Fig. 23.12).

The atomistic modeling technique includes three important categories, namely the molecular dynamics (MD), Monte Carlo (MC), and *ab initio* approaches. The MD and MC simulations are constructed using the second Newton's law and the *ab initio* approach relies on accurate solution of the Schrödinger equation to extract the locations of each atom. Tight bonding molecular dynamics (TBMD), local density approximation (LDA), and density functional theory (DFT) are also another combined approaches, which are often computationally expensive [46]. Although atomistic modeling technique provides a valuable insight into complex structures, due to its huge computational effort especially for large-scale CNTs with high number of walls, its application is limited to the systems with small number of atoms. Therefore, the alternative continuum mechanics approaches were proposed for larger systems or larger time. Figure 23.12 shows the range of different length and time scales with the corresponding theoretical approaches.

Continuum mechanics based approach is considered as an efficient way to save computational resources. This technique employs the continuum mechanics theories of shells, plates, beams, rods and trusses. The continuum mechanics approach by establishing a linkage between structural mechanics and molecular mechanics has aroused widespread interest. Recently, some studies have been developed based on continuum mechanics for estimating elastic properties of CNTs. For instance, Odegard et al.

[165] and Li and Chou [166] developed a molecular structural mechanics approach for modeling CNTs. The finite element (FE) method has been recently introduced to describe the mechanical properties of CNTs. In this regard, Tserpes and Papanikos [153] proposed a FE model for SWNTs. By using a linkage between molecular and continuum mechanics, they determined the elastic moduli of beam element. Furthermore, they reported the dependence of elastic modulus on diameter and chirality of the nanotubes.

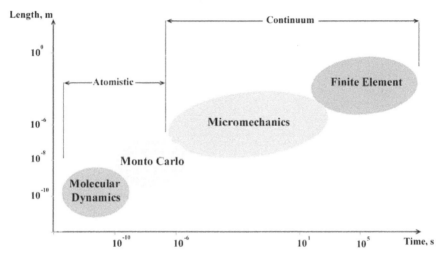

FIGURE 23.12   The range of different length and times scale and theoretical approaches. Adapted from Ref. [46].

It should be noted that the results obtained from these approaches might be higher than those observed in experiments. This may be due to the idealized conditions that are assumed, especially the perfect bonding conditions at the CNT-polymer interfacial region. Table 23.4 shows comparison of experimental and theoretical results for the Young's modulus of SWNT. These theoretical Young's modulus ranges approximately agree with the experimental measurements for the SWNT studied.

TABLE 23.4 Comparison between different theoretical and experimental results of SWNT Young's modulus.

	Method	Wall thickness (nm)	Young's modulus (GPa)	Ref.
	MD	—	1100–1200	[167]
	MD	0.066	5500	[168]
	MD, force approach (FA)	—	1350	[169]
	MD, energy approach (EA)	—	1238	[169]
	ab initio	—	750–1180	[170]
	ab initio	0.089	3859	[171]
	TBMD	0.34	1240	[172]
	TBMD	—	5100	[173]
	DFT	—	1240	[174]
	Molecular mechanics	—	1325	[175]
	MD, Brenner potential	—	5500	[176]
Theoretical approaches	Nanoscale continuum	—	4750–7050	[177]
	LDA combined with elastic shell theory	0.075	4700	[178]
	Molecular structural mechanics model	0.34	Z: 1080±20; A: 1050±20	[179]
	Continuum mechanics model	0.0617	4480	[180]
	Analytical molecular structural mechanics based on the Morse potential	—	1000–1200	[181]
	Energy equivalence between truss and chemical models	—	2520	[182]
	Combined FEM and continuum mechanics, elastic shell	—	4840	[183]
	Structural mechanics approach, space frame, stiffness matrix method	—	1050	[166]
	Molecular mechanics-based FEM, elastic rod element	—	400	[184]
	Space frame and beam	—	9500–1060	[153]
	Beam element and nonlinear spring	—	9700–1050	[185]
	Spring elements	—	1080–1320	[71]
	3-D beam element	—	400–2080	[186]

TABLE 23.4   *(Continued)*

	Method	Wall thickness (nm)	Young's modulus (GPa)	Ref.
	Thermal vibrations-TEM	—	400–4150	[187]
	Cooling-induced vibrations, microRaman spectroscopy	—	2800–3600	[188]
Experimental	Measuring resonance frequency	—	Z: 100–1000	[189]
	Tensile test- AFM	—	1002	[190]
	3-point bending-AFM	—	1200	[191]
	TEM	—	Z: 900–1700	[192]

MD: Molecular dynamics; TBMD: Tight bonding molecular dynamics; DFT: Density functional theory; LDA: Local density approximation; FEM: Finite element method; TEM: Transmission electron microscope; AFM: Atomic force microscope.

[a] Z: Zigzag; B A: Armchair.

23.7.2   SUMMARY OF LITERATURE DATA

Nowadays modeling and simulation of CNTs-polymer composites is of special interest to many researchers. For example, Shokrieh and Rafiee [193] compared the results of finite element (FE) analysis and the rule of mixture and deduced that the latter overestimates the properties of investigated composite and leads to inappropriate results. This observation is in agreement with that of other researchers in literature [194–198]. The effects of CNT aspect ratio, CNT volume fraction, and matrix modulus on axial stress and interfacial shear stress were studied by Haque and Ramasetty [199]. Xiao and Zhang [200]. employed the Cox model to investigate the effects of the length and the diameter of the CNTs on the distributions of the tensile and interfacial shear stress of SWNTs in an epoxy matrix. Eken et al. [201] studied the combined effects of nanotube aspect ratio and shear rate on the electrical properties and microstructure of CNT-polymer composite. They presented that the orientation of the nanotubes and electrical properties of the CNT-polymer composites strongly depend on the nanotube aspect ratio and applied shear rate. Georgantzinos et al. [202] carried out multiscale FE formulation to investigate the stress-strain behavior of rubber uniformly filled with continuous SWNTs. They concluded that the SWNTs improve significantly the composite strength and toughness especially for higher volume fractions. A molecular structural mechanics model of CNT-reinforced composites has been proposed by Wang et al. [203]. It is assumed that CNTs and matrix are perfectly bonded to each other. They also presented the effects of matrix materials, van der Waals force, chiral vector, and tube layers of CNTs on interfacial stresses in CNT-reinforced composites. In the study on multiscale modeling of CNT-reinforced nanofiber scaffold by Unnikrishnan et al., [204] it was demonstrated that the CNT polymeric scaffold properties estimated using both the micromechanical and nonlinear hyperelastic material homogenization models show excellent comparison with experimental data. In another study, mechanical performance of CNT-reinforced polymer composites at cryogenic temperatures was performed experimentally and numerically by Takeda et al. [205]. FE computations were used to model the representative volume element (RVE) of the composites, and the possible existence of the imperfect CNT-polymer interface bonding was considered. Joshi et al. [164] implemented the continuum mechanics approach using a hexagonal RVE under lateral loading conditions to investigate the mechanical properties of CNT reinforced composite. They also used an extended rule of mixtures (ERM) to validate the proposed model. The reported theoretical approaches on mechanical properties of CNT-polymer composites including corresponding validation technique and interface bonding are given in Table 23.5.

TABLE 23.5 Theoretical approaches on predicting the mechanical properties of CNT-polymer composite [46].

Theoretical approaches	Validation technique	SWNT in matrix		Interface bonding
		Single	Disperse	
Equivalent continuum mechanics	Mechanical testing	×		Perfect
FEM, strain energy method	Literature data	×		Perfect
MD, energy minimization method	Rule of mixture		×	Perfect
MD, Brenner and united-atom potential	Ordinary and extended rule of mixture		×	Perfect

TABLE 23.5   *(Continued)*

Theoretical approaches	Validation technique	SWNT in matrix		Interface bonding
		Single	Disperse	
Combined FEM and microme-chanics models (wavy and straight SWNTs)	Literature data	×		Perfect
Square RVE, FEM, rule of mix-ture, and elasticity theory	Consistent with some experimental data	×		Perfect
Micromechanics equations for randomly oriented short fibers	Experimental	×		Perfect
Multiscale Monte Carlo FEM using the equivalent continuum method	Tensile test at the macroscale	×		Perfect
FEM, strain energy method	Literature data	×		Perfect
Beam elements for SWCNTs us-ing molecular structural mechan-ics, truss rod for vdW links, cubic elements for the matrix	—		×	vdW interaction

vdW: van der Waals.

23.8   CONCLUDING REMARKS AND FUTURE WORK

CNTs have emerged as the promising reinforcements for polymer com-posites due to their high aspect ratios and exceptionally remarkable me-chanical properties. Studies have shown that the mechanical properties of CNT-polymer composites affected by many parameters, including type of CNTs, chirality, purity, aspect ratio, loading content, dispersion, alignment and interfacial adhesion between the CNT and polymer matrix. Among them, the interfacial adhesion is identified as one of the most important factors affecting the performance of nanocomposites.

The introduction of CNTs into electrospun polymer nanofibers is an interesting attempt to enhance the mechanical properties of electrospun CNT-polymer composite nanofibers. So far, a few reports have been found in the literature regarding measurement of the mechanical properties of single electrospun composite nanofiber. Isolation of individual composite nanofiber remains as the biggest challenge in measuring tensile properties of composite nanofiber, due to the small size of the samples. Another challenge is to grip it into a sufficiently small load scale tester. Although a number of studies have been performed on the experimental investigation of electrospun CNT-polymer composite, more attention should be given to the theoretical approaches on predicting the mechanical properties of CNTs embedded in electrospun fiber mat. Accordingly, this review summarizes recent results on modeling and simulation of CNT-polymer composites.

Along with the development of theoretical and computational approaches for nanotubes, many techniques have been demonstrated in many studies using atomistic modeling and continuum mechanics approaches. A molecular simulation through atomistic approach, such as molecular dynamics or Monte Carlo, has been successfully used in material science and engineering to determine the physical and mechanical properties of materials in nanoscale. One of the main challenges in applying such a simulation to a single electrospun composite nanofiber is the huge amount of computational tasks especially for large number of atoms. Another challenge is to determine the structure and arrangement of nanofiber and also to establish an interatomic potential function.

KEYWORDS

- carbon nanotube
- electrospinning
- mechanical properties
- theoretical approaches

REFERENCES

1. Haghi, A. K., Zaikov, G., Advances in Nanofiber Research, *i*Smithers, UK, 2011
2. Fang, J., Wang, X., Lin, T., in "Nanofibers-Production, Properties and Functional Applications," InTech, 2011.
3. Li, D., Xia, Y., *Adv. Mater.*, **16**, 1151 (2004).
4. Reneker, D. H., , Chun, I., *Nanotechnology*, **7**, 216 (1996).

5. Baji, A., Mai, Y. W., Wong, S. C., Abtahi, M., Chen, P., *Compos. Sci. Technol.*, **70**, 703 (2010).
6. Mottaghitalab, V., Haghi, A. K., *Korean. J. Chem. Eng.*, **28**, 114 (2011).
7. Ziabari, M., Mottaghitalab, V., A.K. Haghi in "Nanofibers: Fabrication, Performance, and Applications," Nova Science Publishers, New York, 2009.
8. Shenoy, S. L., Bates, W. D., Frisch, H. L., Wnek, G. E., *Polymer*, **46**, 3372 (2005).
9. Burger, C., Hsiao, B. S., Chu, B., *Annu. Rev. Mater. Res.*, **36**, 333 (2006).
10. Haghi, A. K., Akbari, M., *Phys. Status. Solidi. A.*, **204**, 1830 (2007).
11. Nasouri, K., Bahrambeygi, H., Rabbi, A., Shoushtari, A. M., Kaflou, A., *J. Appl. Polym. Sci.*, **126**, 127 (2012).
12. Subbiah, T., Bhat, G. S., Tock, R. W., Parameswaran, S., Ramkumar, S. S., *J. Appl. Polym. Sci.*, **96**, 557 (2005).
13. Hasanzadeh, M., B. Hadavi Moghadam, in "Research Progress in Nanoscience and Nanotechnology," Nova Science Publisher, New York, 2012.
14. Park, J. H., Kim, B. S., Yoo, Y. C., Khil, M. S., Kim, H. Y., *J. Appl. Polym. Sci.*, **107**, 2211 (2008).
15. Zhou, F. L., Gong, R. H., *Polym. Int.*, **57**, 837 (2008).
16. Nisbet, D. R., Forsythe, J. S., Shen, W., Finkelstein, D. I., Horne, M. K., *J. Biomater. Appl.*, **24**, 7 (2009).
17. Bhattarai, S. R., Bhattarai, N., Yi, H. K., Hwang, P. H., Cha, D. I., Kim, H. Y., *Biomaterials*, **25**, 2595 (2004).
18. Li, W. J., Laurencin, C. T., Caterson, E. J., Tuan, R. S., Ko, F. K., *J. Biomed. Mater. Res.*, **60**, 613 (2002).
19. Lu, H., Chen, W., Xing, Y., Ying, D., Jiang, B., *J. Bioact. Compat. Pol.*, **24**,158 (2009)
20. Lee, S., Obendorf, S. K., *Text. Res. J.*, **77**, 696 (2007).
21. Sill, T. L., H.A. von Recum, *Biomaterials*, **29**, 1989 (2008).
22. Kim, C., *J. Power. Sources*, **142**, 382 (2005).
23. Kim, C., B.Ngoc, T. N., Yang, K. S., Kojima, M., Kim, Y. A., Kim, Y. J., Endo, M., Yang, S. C., *Adv. Mater.*, **19**, 2141 (2007).
24. Thavasi, V., Singh, G., Ramakrishna, S., *Energy. Environ. Sci.*, **1**, 205 (2008).
25. Salvetat, J. P., G.Briggs, A. D., Bonard, J. M., Bacsa, R. R., Kulik, A. J., Stöckli, T., Burnham, N. A., Forró, L., *Phys. Rev. Lett.*, **82**, 944 (1999).
26. Ebbesen, T. W., Lezec, H. J., Hiura, H., Bennett, J. W., Ghaemi, H. F., Thio, T., *Nature*, **382**, 54 (1996).
27. Berber, S., Kwon, Y. K., Tomanek, D., *Phys. Rev. Lett.*, **84**, 4613 (2000).
28. Dai, H., *Surface. Sci.*, **500**, 218 (2002).
29. I. Kanga, Y.Y. Heung, J.H. Kim, J.W. Lee, R. Gollapudi, S. Subramaniam, S. Narasimhadevara,, G.R. Kirikera, V. Shanov, M.J. Schulz, D. Shi, J. Boerio, S. Mall, M. Ruggles-Wren, *Compos. Part B*, **37**, 382 (2006).
30. Salvetat-Delmotte, J. P., Rubio, A., *Carbon*, **40**, 1729 (2002).
31. Naebe, M., Lin, T., Wang, X., in "Nanofibers," InTech, 2010.
32. Choudhary, V., Gupta, A., in "Carbon Nanotubes-Polymer Nanocomposites," InTech, 2011.
33. Gao, J., Yu, A., Itkis, M. E., Bekyarova, E., Zhao, B., Niyogi, S., *J. Am. Chem. Soc.*, 126, 16698 (2004).
34. Chronakis, I. S., *J. Mater. Process. Tech.*, **167**, 283 (2005).

35. Ko, F., Gogotsi, Y., Ali, A., Naguib, N., Ye, H., Yang, G., Li, G., Willis, P., *Adv. Mater.*, **15**, 1161 (2003).
36. Naebe, M., Lin, T., Tian, W., Dai, L., Wang, X., *Nanotechnology*, **18**, 1 (2007).
37. Naebe, M., Lin, T., Staiger, M. P., Dai, L., Wang, X., *Nanotechnology*, **19**, 305702 (2008).
38. Mathew, G., Hong, J. P., Rhee, J. M., Lee, H. S., Nah, C., *Polym. Test.*, **24**, 712 (2005).
39. Nasouri, K., Shoushtari, A. M., Kaflou, A., Bahrambeygi, H., Rabbi, A., *Polym. Composite*, **33**, 1951.
40. Mahdieh, Z. M., Mottaghitalab, V., Piri, N., Haghi, A. K., *Korean. J. Chem. Eng.*, **29**, 111 (2012).
41. Mottaghitalab, V., Haghi, A. K., in "Development of Nanotechnology in Textiles," Nova Science Publisher, New York, 2012.
42. Mazinani, S., Ajji, A., Dubois, C., *Polymer*, **50**, 3329 (2009).
43. Islam, M. S., Ashaduzzaman, M., Masum, S. M., Yeum, J. H., *Dhaka. Univ. J. Sci.*, **60**, 125 (2012).
44. Zhang, Q., Chang, Z., Zhu, M., Mo, X., Chen, D., *Nanotechnology*, **18**, 115611 (2007).
45. Ye, H., Lam, H., Titchenal, N., Y. Gogotsi, Ko, F., *Appl. Phys. Lett.*, **85**, 1775 (2004).
46. Shokrieh, M. M., Rafiee, R., *Mech. Compos. Mater.*, **46**, 155 (2010).
47. Kilic, A., Oruc, F., Demir, A., *Text. Res. J.*, **78**, 532 (2008).
48. A. Shams Nateri, Hasanzadeh, M., *J. Comput. Theor. Nanosci.*, **6**, 1542 (2009).
49. Ziabari, M., Mottaghitalab, V., Haghi, A. K., *Korean. J. Chem. Eng.*, **27**, 340 (2010).
50. Ziabari, M., Mottaghitalab, V., Haghi, A. K., *Braz. J. Chem. Eng.*, **26**, 53 (2009)
51. Ziabari, M., Mottaghitalab, V., McGovern, S. T., Haghi, A. K., *Chin. Phys. Lett.*, **25**, 3071 (2008).
52. Ziabari, M., Mottaghitalab, V., McGovern, S. T., Haghi, A. K., *Nanoscale. Res. Lett.*, **2**, 297 (2007).
53. Ziabari, M., Mottaghitalab, V., Haghi, A. K., *Korean. J. Chem. Eng.*, **25**, 919 (2008).
54. Ziabari, M., Mottaghitalab, V., Haghi, A. K., *Korean. J. Chem. Eng.*, **25**, 905 (2008).
55. Ziabari, M., Mottaghitalab, V., Haghi, A. K., *Korean. J. Chem. Eng.*, **25**, 923 (2008).
56. Nasouri, K., Shoushtari, A. M., Kaflou, A., *Micro. Nano. Lett.*, **7**, 423 (2012).
57. Sabetzadeh, N., Bahrambeygi, H., Rabbi, A., Nasouri, K., *Micro. Nano. Lett.*, **7**, 662 (2012).
58. Yarin, A. L., Koombhongse, S., Reneker, D. H., *J. Appl. Phys.*, **90**, 4836 (2001).
59. Doshi, J., Reneker, D. H., *J. Electrostat.*, **35**, 151 (1995).
60. Theron, S. A., Zussman, E., Yarin, A. L., *Polymer*, **45**, 2017 (2004).
61. Reneker, D. H., Yarin, A. L., Fong, H., Koombhongse, S., *J. Appl. Phys.*, **87**, 4531 (2000).
62. Feng, J. J., *Phys. Fluids.*, **14**, 3912 (2002).
63. Carroll, C. P., Joo, Y. L., *Phys. Fluids.*, **18**, 053102 (2006)
64. Zhmayev, E., Zhou, H., Joo, Y. L., *J. Non-Newton. Fluid. Mech.*, **153**, 95 (2008)
65. Stanger, J. J., M. Sc. thesis, University of Canterbury, 2008.
66. Shin, Y. M., Hohman, M. M., Brenner, M. P., Rutledge, G. C., *Polymer*, **42**, 9955 (2001)
67. Zhang, S., Shim, W. S., Kim, J., *Mater. Design.*, **30**, 3659 (2009).
68. Yördem, O. S., Papila, M., Menceloğlu, Y. Z., *Mater. Design.*, **29**, 34 (2008).
69. Agic, A., *J. Appl. Polym. Sci.*, **108**, 1191 (2008).

70. Meo, M., Rossi, M., *Compos. Sci. Technol.*, **66**, 1597 (2006).
71. Giannopoulos, G. I., Kakavas, P. A., Anifantis, N. K., *Comp. Mater. Sci.*, 41, 561 (2008).
72. Natsuki, T., Tantrakarn, K., Endo, M., *Appl. Phys. A.*, **79**, 117 (2004).
73. Ansari, R., Rouhi, S., *Physica. E.*, **43**, 58 (2010).
74. Qian, D., Wagner, G. J., Liu, W. K., Yu, M. F., Ruoff, R. S., *Appl. Mech. Rev.*, 55, 495 (2002).
75. Wernik, J. M., Meguid, S. A., *Appl. Mech. Rev.*, **63**, 050801 (2010).
76. Belin, T., Epron, F., *Mater. Sci. Eng. B.*, **119**, 105 (2005).
77. Popov, V. N., *Mater. Sci. Eng. R.*, **43**, 61 (2004).
78. Al-Saleh, M. H., Sundararaj, U., *Carbon*, **47**, 2 (2009).
79. Ma, P. C., Siddiqui, N. A., Marom, G., Kim, J. K., *Composites. A.*, **41**, 13451367 (2010)
80. Al-Saleh, M. H., Sundararaj, U., *Composites. A.*, **42**, 2126 (2011).
81. Coleman, J. N., Khan, U., Blau, W. J., Y.K. Gun'ko, *Carbon*, **44**, 1624 (2006).
82. Moridi, Z., Mottaghitalab, V., Haghi, A. K., *Cellulose. Chem. Technol.*, **45**, 549 (2011).
83. Khare, R., Bose, S., *J. Miner. Mater. Charact. Eng.*, **4**, 31 (2005).
84. Chiolerio, A., Castellino, M., Jagdale, P., Giorcelli, M., Bianco, S., Tagliaferro, A., in "Carbon Nanotubes-Polymer Nanocomposites," InTech, 2011.
85. Breuer, O., Sundararaj, U., *Polym. Composite.*, **25**, 630 (2004).
86. Min, B. G., Chae, H. G., Minus, M. L., Kumar, S., in "Functional Composites of Carbon Nanotubes and Applications," Transworld Research Network, India, 2009.
87. A. Agic A, Mijovic, B., in "Engineering the Future," InTech, 2010.
88. Dror, Y., Salalha, W., Khalfin, R. L., Cohen, Y., Yarin, A. L., Zussman, E., *Langmuir*, **19**, 7012 (2003).
89. Yeo, L. Y., Friend, J. R., *J. Exp. Nanosci.*, **1**, 177 (2006).
90. K.Wong, K. H., Zinke-Allmanga, M., Hutter, J. L., Hrapovic, S., J.Luong, H. T., Wan, W., *Carbon*, **47**, 2571 (2009).
91. Zhou, W., Wu, Y., Wei, F., Luo, G., Qian, W., *Polymer*, **46**, 12689 (2005).
92. Tiwari, D. C., Sen, V., Sharma, R., *Indian. J. Pure. Ap. Phy.*, **50**, 49 (2012).
93. Heikkilä, P., Harlin, A., *eXPRESS Polym. Lett.*, **3**, 437 (2009).
94. Ra, E. J., An, K. H., Kim, K. K., Jeong, S. Y., Lee, Y. H., *Chem. Phys. Lett.*, **413**, 188 (2005)
95. Wan, Y. Q., He, J. H., Yu, J. Y., *Polym. Int.*, **56**, 1367 (2007).
96. Vaisman, L., Wachtel, E., Wagner, H. D., Marom, G., *Polymer*, 48, 6843 (2007).
97. Uddin, N. M., Ko, F., Xiong, J., Farouk, B., Capaldi, F., *Res. Lett. Mater. Sci.*, Doi: 10.1155/2009/868917 (2009).
98. Qiao, B., Ding, X., Hou, X., Wu, S., *J. Nanomater.*, Doi: 10.1155/2011/839462, (2011).
99. Ji, J., Sui, G., Yu, Y., Liu, Y., Lin, Y., Du, Z., Ryu, S., Yang, X., *J. Phys. Chem. C.*, **113**, 4779 (2009)
100. Liu, L. Q., Tasis, D., Prato, M., Wagner, H. D., *Adv. Mater.*, **19**, 1228 (2007).
101. Sundaray, B., Babu, V. J., Subramanian, V., Natarajan, T. S., *J. Eng. Fiber. Fabr.*, **3**, 39 (2008).
102. Shao, S., Li, L., Yang, G., Li, J., Luo, C., Gong, T., Zhou, S., *Int. J. Pharmaceut.*, **421**, 310 (2011).
103. Pan, C., Ge, L. Q., Gu, Z. Z., *Compos. Sci. Technol.*, **67**, 3271 (2007).

104. Shao, S., Zhou, S., Li, I., Li, J., Luo, C., Wang, J., Li, X., Weng, J., *Biomaterials*, **32**, 2821 (2011).
105. Liao, G. Y., Zhou, X. P., Chen, L., Zeng, X. Y., Xie, X. L., Mai, Y. W., *Compos. Sci. Technol.*, **72**, 248 (2012).
106. Saeed, K., Park, S. Y., Lee, H. J., Baek, J. B., Huh, W. S., *Polymer*, **47**, 8019 (2006).
107. Kim, G. M., Michler, G. H., P. Pötschke, *Polymer*, **46**, 7346 (2005).
108. Ahn, B. W., Chi, Y. S., Kang, T. J., *J. Appl. Polym. Sci.*, **110**, 4055 (2008).
109. Baji, A., Mai, Y. W., Wong, S. C., *Mater. Sci. Eng. A.*, **528**, 6565 (2011).
110. Ketpang, K., Park, J. S., *Synthetic. Met.*, **160**, 1603 (2010).
111. Im, J. S., Kim, J. G., Lee, S. H., Lee, Y. S., *Colloid. Surface. A: Physicochem. Eng. Aspects.*, **364**, 151 (2010).
112. Shin, M. K., Kim, Y. J., Kim, S. I., Kim, S. K., Lee, H., Spinks, G. M., Kim, S. J., *Sensor. Actuat. B.*, **134**, 122 (2008).
113. Han, G., Shi, G., *J. Appl. Polym. Sci.*, **103**, 1490 (2007).
114. Mathew, G., Hong, J. P., Rhee, J. M., Lee, H. S., Nah, C., *Polym. Test.*, **24**, 712 (2005).
115. Konkhlang, T., Tashiro, K., Kotaki, M., Chirachanchai, S., *J. Am. Chem. Soc.*, **130**, 15460 (2008).
116. Coleman, J. N., Khan, U., Y.K. Gun'ko, *Adv. Mater.*, **18**, 689 (2006).
117. Schadler, L. S., Giannaris, S. C., Ajayan, P. M., *Appl. Phys. Lett.*, **73**, 3842 (1998).
118. Larijani, M. M., Khamse, E. J., Asadollahi, Z., Asadi, M., *Bull. Mater. Sci.*, **32**, 305 (2012).
119. Wang, Q., Dai, J., Li, W., Wei, Z., Jiang, J., *Compos. Sci. Technol.*, **68**, 1644 (2008).
120. Xie, X. L., Mai, Y. W., Zhou, X. P., *Mater. Sci. Eng.*, **49**, 89 (2005).
121. Hassan, M. A., *Al-Qadisiya. J. Eng. Sci.*, **5**, 341 (2012).
122. Bal, S., Samal, S. S., *Bull. Mater. Sci.*, **30**, 379 (2007).
123. Xu, L. R., Sengupta, S., *J. Nanosci. Nanotechnol.*, **5**, 620 (2005).
124. Pandurangappa, M., Raghu, G. K., in "Carbon Nanotubes Applications on Electron Devices," InTech, 2011.
125. Ham, H. T., Koo, C. M., Kim, S. O., Choi, Y. S., Chung, I. J., *Macromol. Res.*, **12**, 384 (2004).
126. Abuilaiwi, F. A., Laoui, T., Al-Harthi, M., Atieh, M. A., *Arab. J. Sci. Eng.*, **35**, 37 (2010).
127. Andrews, R., Weisenberger, M. C., *Curr. Opin. Solid. State. Mat. Sci.*, **8**, 31 (2004).
128. Chew, S. Y., Hufnagel, T. C., Lim, C. T., Leong, K. W., *Nanotechnology*, **17**, 3880 (2006).
129. Gu, S. Y., Wu, Q. L., Ren, J., Vancso, G. J., *Macromol. Rapid. Comm.*, **26**, 716 (2005).
130. E.Tan, P. S., Goh, C. N., Sow, C. H., Lim, C. T., *Appl. Phys. Lett.*, **86**, 073115 (2005).
131. L. Yang,, K.O. van der Werf, M.L. Bennink, P.J. Dijkstra, J. Feijen, *Biomaterials*, **29**, 955 (2008).
132. Almecija, D., Blond, D., Sader, J. E., Coleman, J. N., Boland, J. J., *Carbon*, **47**, 2253 (2009).
133. E.Tan, P. S., Lim, C. T., *Compos. Sci. Technol.*, **66**, 1102 (2006).
134. Gandhi, M., Yang, H., Shor, L., Ko, F., *Polymer*, **50**, 1918 (2009).
135. Huang, Z. M., Zhang, Y. Z., Kotaki, M., Ramakrishna, S., *Compos. Sci. Technol.*, **63**, 2223 (2003).

136. Pham, Q. P., Shamra, U., Mikos, A. G., *Tissue. Eng.*, **12**, 1197 (2006).
137. Andrady, A. L., "Science and technology of polymer nanofibers," Wiley, Canada, 2008.
138. He, J. H., Liu, Y., Mo, L. F., Wan, Y. Q., Xu, L., "Electrospun Nanofibers and Their Applications," *i*Smithers, UK, 2008.
139. Ramakrishna, S., Fujihara, K., Teo, W. E., Lim, T. C., Ma, Z., "An introduction to electrospinning and nanofibers," World Scientific Publishing, Singapore, 2005.
140. Brown, P. J., Stevens, K., "Nanofibers and nanotechnology in textiles," Woodhead, England, 2007.
141. Schreuder-Gibson, H. L., Gibson, P., Senecal, K., Sennett, M., Walker, J., Yeomans, W., Ziegler, D., Tsai, P. P., *J. Adv. Mater.*, **34**, 44 (2002).
142. Fang, J., Niu, H. T., Lin, T., Wang, W. G., *Chinese. Sci. Bull.*, **53**, 2265 (2008).
143. Kim, D. K., Park, S. H., Kim, B. C., Chin, B. D., Jo, S. M., Kim, D. Y., *Macromol. Res.*, **13**, 521 (2005).
144. Ma, Z., Kotaki, M., Yong, T., He, W., Ramakrishna, S., *Biomaterials*, **26**, 2527 (2005).
145. Lee, B. O., Woo, W. J., Kim, M. S., *Macromol. Mater. Eng.*, **286**, 114 (2001).
146. Wang, Z. G., Xu, Z. K., Wan, L. S., Wu, J., *Macromol. Rapid. Commun.*, **27**, 516 (2006).
147. Shokrgozar, M. A., Mottaghitalab, F., Mottaghitalab, V., Farokhi, M., *J. Biomed. Nanotechnol.*, **7**, 276 (2011).
148. J.S. Im, J.G. Kim, S.H Lee, Y.S. Lee, *Colloid. Surface. A: Physicochem. Eng. Aspects.*, **364**, 151 (2010).
149. Q. Jiang, G. Fu, D. Xie, S. Jiang, Z. Chen, B. Huang, Y. Zhao, *Proced. Eng.*, **27**, 72 (2012).
150. Wanna, Y., Pratontep, S., Wisitsoraat, A., Tuantranont, A., 5th IEEE conference on sensor, 342–345, 2006.
151. Abdullah, S. A., Frormann, L., in "Nanofibers: Fabrication, Performance, and Applications," Nova Science Publishers, New York, 2009.
152. Shokrieh, M. M., Rafiee, R., *Mater. Design.*, **31**, 790 (2010).
153. Tserpes, K. I., Papanikos, P., *Compos. Part. B.*, **36**, 468 (2005).
154. Arani, A. G., Rahmani, R., Arefmanesh, A., *Physica E.*, **40**, 2390 (2008).
155. Ruoff, R. S., Qian, D., Liu, W. K., *C R Physique*, **4**, 993 (2003).
156. Guo, X., A.Leung, Y. T., He, X. Q., Jiang, H., Huang, Y., *Compos. Part B.*, **39**, 202 (2008).
157. Xiao, J. R., Gama, B. A., J.W. Gillespie Jr, *Intl. J. Solids. Struct.*, **42**, 3075 (2005).
158. Ansari, R., Motevalli, B., *Commun. Nonlinear. Sci. Numer. Simulat.*, **14**, 4246 (2009).
159. Li, C., Chou, T. W., *Mech. Mater.*, **36**, 1047 (2004).
160. Natsuki, T., Endo, M., *Carbon*, **42**, 2147 (2004).
161. Ansari, R., Sadeghi, F., Motevalli, B., *Commun. Nonlinear. Sci. Numer. Simulat.*, **18**, 769 (2013).
162. Alisafaei, F., Ansari, R., *Comp. Mater. Sci.*, **50**, 1406 (2011).
163. Natsuki, T., Ni, Q. Q., Endo, M., *Appl. Phys. A.*, **90**, 441 (2008).
164. Joshi, U. A., Sharma, S. C., Harsha, S. P., *Proc. IMechE., Part N: J. Nanoengineering. Nanosystems.*, **225**, 23 (2011).
165. Odegard, G. M., Gates, T. S., Nicholson, L. M., Wise, K. E., *Compos. Sci. Technol.*, **62**, 1869 (2002).

166. Li, C., Chou, T. W., *Int. J. Solids. Struct.*, **40**, 2487 (2003).
167. Prylutskyy, Y. I., Durov, S. S., Ogloblya, O. V., Buzaneva, E. V., Scharff, P., *Comp. Mater. Sci.*, **17**, 352 (2000).
168. Yakobson, B. I., Brabec, C. J., Bernholc, J., *Phys. Rev. Lett.*, **76**, 2511 (1996).
169. Jin, Y., Yuan, F. G., *Compos. Sci. Technol.*, **63**, 1507 (2003).
170. Lier, G. V., Alsenoy, C. V., Doren, V. V., Geerlings, P., *Chem. Phys. Lett.*, **326**, 181 (2000)
171. Kudin, K. N., Scuseria, G. E., Yakobson, B. I., *Phys. Rev. B.*, **64**, 235406 (2001).
172. Hernandez, E., Goze, C., Bernier, P., Rubio, A., *Phys. Rev. Lett.*, **80**, 4502 (1998).
173. Xin, Z., Jianjun, Z., Zhong-can, O. Y., *Phys. Rev. B.*, **62**, 13692 (2000).
174. Hernandez, E., Goze, C., Bernier, P., Rubio, A., *Appl. Phys. A.*, **68**, 287 (1999).
175.   Chang, T., Gao, H., *J. Mech. Phys. Solids.*, **51**, 1059 (2003).
176. Yakobson, B. I., Brabec, C. J., Bernholc, J., *Phys. Rev. Lett.*, **76**, 2511 (1996).
177. Zhang, P., Huang, Y., Geubelle, P. H., Klein, P. A., Hwang, K. C., *Int. J. Solids. Struct.*, **39**, 3893 (2002).
178. Tu, Z., Ou-Yang, Z., *Phys. Rev. B.*, **65**, 233407 (2002).
179. Li, C. Y., Chou, T. W., *Compos. Sci. Technol.*, **63**, 1517 (2003).
180. Vodenitcharova, T., Zhang, L. C., *Phys. Rev. B.*, **68**, 165401 (2003).
181. Wang, Q., *Int. J. Solids. Struct.*, **41**, 5451 (2004).
182. Selmi, A., Friebel, C., Doghri, I., Hassis, H., *Compos. Sci. Technol.*, **67**, 2071 (2007).
183. Pantano, A., Parks, D. M., Boyce, M. C., *J. Mech. Phys. Solids.*, **52**, 789 (2004).
184. Sun, X., Zhao, W., *Mater. Sci. Eng. A.*, **390**, 366 (2005).
185. Kalamkarov, A. L., Georgiades, A. V., Rokkam, S. K., Veedu, V. P., GhasemiNejhad, M. N., *Int. J. Solids. Struct.*, **43**, 6832 (2006).
186. Papanikos, P., Nikolopoulos, D. D., Tserpes, K. I., *Comput. Mater. Sci.*, **43**, 345 (2008).
187. Treacy, M. M., Ebbesen, T. W., Gibson, J. M., *Nature*, **38**, 678 (1996).
188. Lourie, O., Cox, D. M., Wagner, H. D., *Phys. Rev. Lett.*, **81**, 1638 (1998).
189. Liu, J. Z., Zheng, Q. S., Jiang, Q., *Phys. Rev. Lett.*, **86**, 4843 (2001).
190. Yu, M. F., Lourie, O., Dyer, M. J., Moloni, K., Kelly, T. F., Ruo, R. S., *Science*, **287**, 637 (2000)
191. Tombler, T. W., Zhou, C., Kong, J., Dai, H., Liu, L., Jayanthi, C. S., Tang, M., Wu, S. Y., *Nature*, **405**, 769 (2000).
192. Krishnan, A., Dujardin, E., Ebbesen, T. W., Yianilos, P. N., M.Treacy, M. J., *Phys. Rev. B.*, **58**, 14013 (1998).
193. Shokrieh, M. M., Rafiee, R., *Mech. Res. Commun.*, **37**, 235 (2010).
194. Tserpes, K. I., Papanikos, P., Labeas, G., Sp.Pantelakis, G., *Theor. Appl. Fract. Mec.*, **49**, 51 (2008).
195. Villoria, R. G., Miravete, A., *Acta. Mater.*, **55**, 3025 (2007).
196. S.Frankland, J. V., Harik, V. M., Odegard, G. M., Brenner, D. W., Gates, T. S., *Compos. Sci. Technol.*, **63**, 1655 (2003).
197. Odegard, G. M., Gates, T. S., Wise, K. E., Parka, C., Siochi, E. J., *Compos. Sci. Technol.*, **63**, 1671 (2003).
198. Saffar, K., P. A., Pour, N. J., Najafi, A. R., Rouhi, G., Arshi, A. R., Fereidoon, A., *World. Acad. Sci. Eng. Tech.*, **47**, 219 (2008).
199. Haque, A., Ramasetty, A., *Compos. Struct.*, **71**, 68 (2005).

200. Xiao, K. Q., Zhang, L. C., *J. Mater. Sci.*, **39**, 4481 (2004).
201. Eken, A. E., Tozzi, E. J., Klingenberg, D. J., Bauhofer, W., *Polymer*, **53**, 4493 (2012).
202. Georgantzinos, S. K., Giannopoulos, G. I., Anifantis, N. K., *Theor. Appl. Fract. Mec.*, **52**, 158 (2009).
203. Wang, H., Meng, F., Wang, X., *J. Reinf. Plast. Comp.*, **29**, 2262 (2010).
204. Unnikrishnan, V. U., Unnikrishnan, G. U., Reddy, J. N., *Compos. Struct.*, **93**, 1008 (2011)
205. Takeda, T., Shindo, Y., Narita, F., Mito, Y., *Mater. Trans.*, **50**, 436 (2009).

INDEX